The Light of Alexandria

Hypatia of Alexandria, the last director of the greatest library of the ancient world

Public domain photo

The Light of Alexandria

By James Maynard

Independently published through

Lulu (www.Lulu.com)

First published in the United States of America
By James Maynard,
Keene, New Hampshire, USA.

First Edition.

FRONT COVER: Discussion was an integral part of the growth of science in
Ancient Ionia and Greece. Here we see two Greek citizens debating a topic,
etched into the wall of a Grecian building.
Rights for commercial use purchased by James Maynard.

REAR COVER: During the time of the growth of science in the ancient world,
art, music and architecture also flourished. This statue is just one example of the
beauty of ancient Greek art.
Rights for commercial use purchased by James Maynard.
Author photograph by Leona Swett.
Photo courtesy of artist.

For my father, Robert Maynard,
author of "Bright and Shining"

Map of the Aegean world

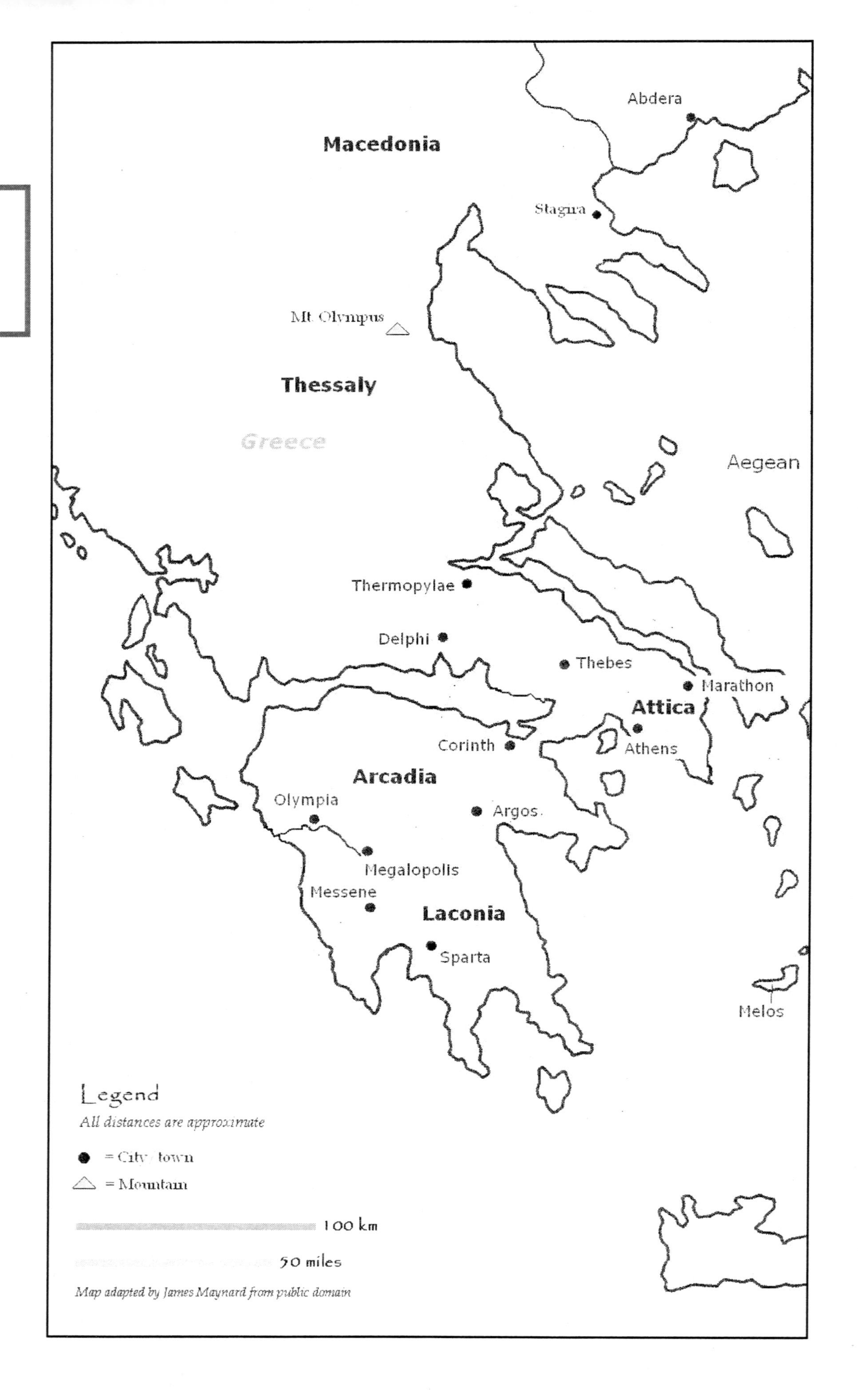

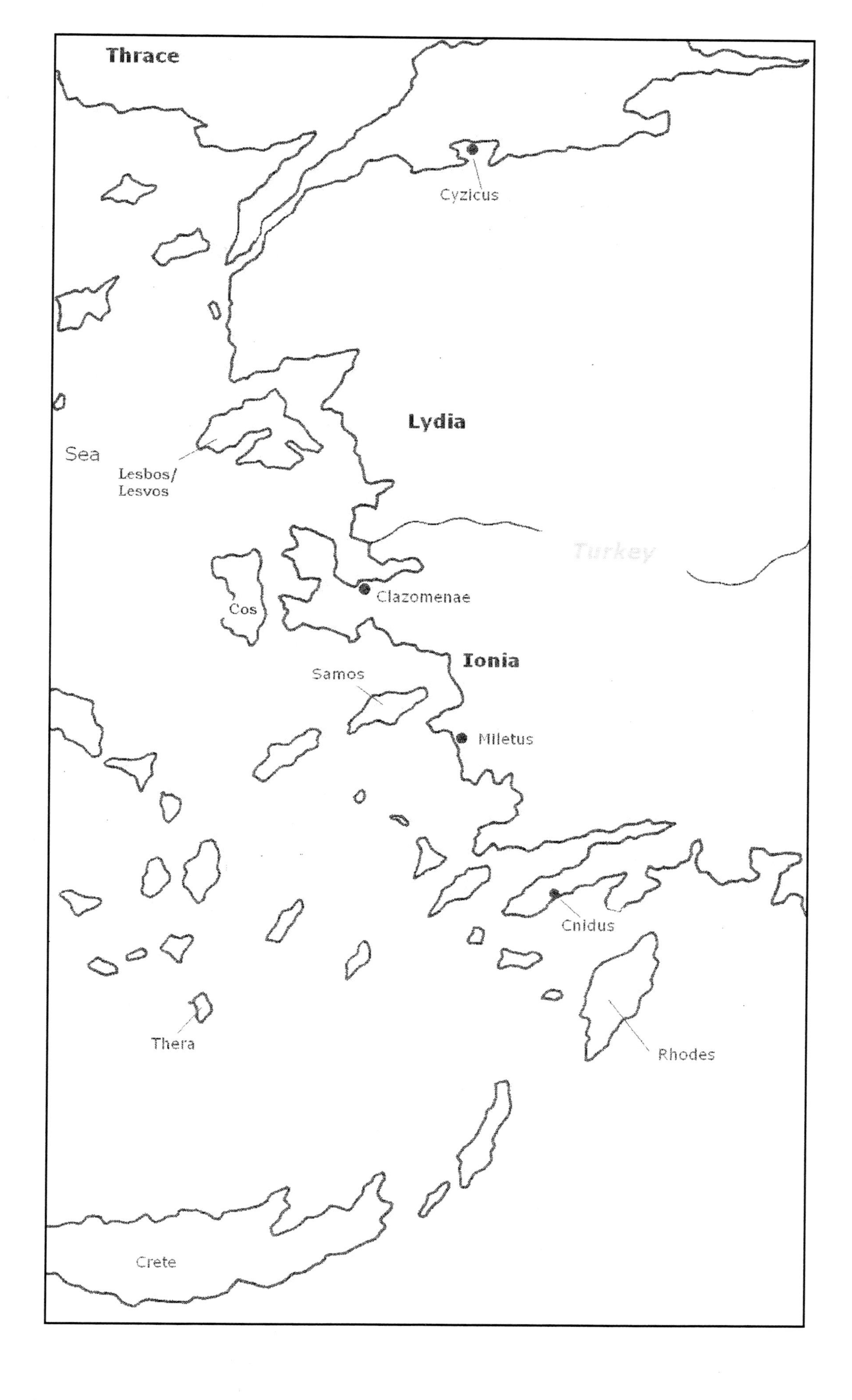
Thrace
Cyzicus
Lydia
Sea
Lesbos/
Lesvos
Turkey
Cos
Clazomenae
Ionia
Samos
Miletus
Cnidus
Thera
Rhodes
Crete

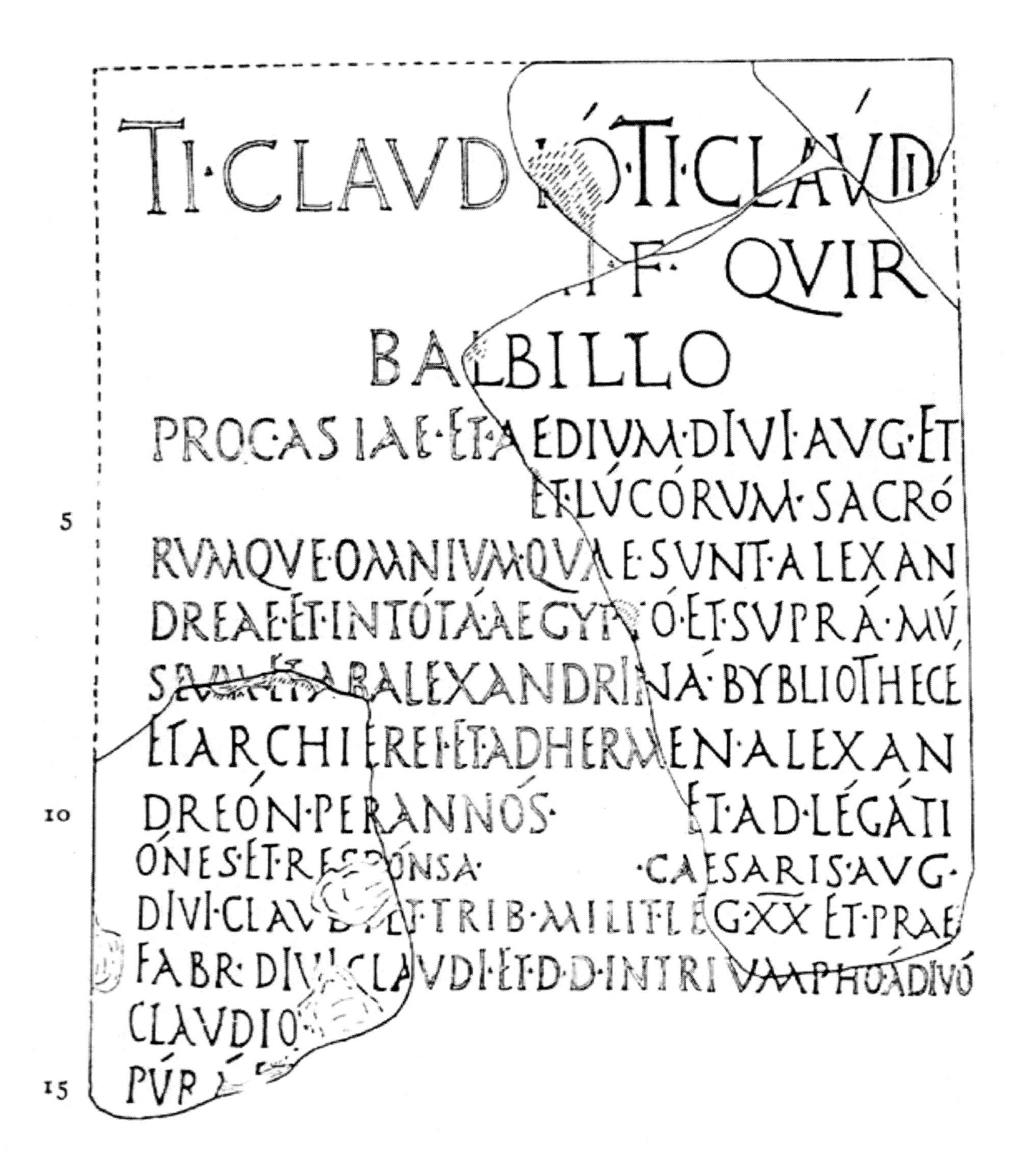

An inscription from the First century CE about operations at the Great Library of Alexandria. The library was already well into decline at this point in history.

Public domain photo

Table of contents

Preface

Science is more than just a collection of ideas. It also represents a world-view. It is a paradigm to represent one's belief of the past, present and future, from the birth of the universe until the end of time.

Modern science was born in Ionia, in the seventh century BCE, for a number of different reasons, some overt and others subtle. Everything from the nature of islands themselves creating maritime societies, to the diversity of immigrant populations and the life cycle of the olive all played a part in why science was first born.

In *The Light of Alexandria*, I hope to present a summary of the forces that gave birth to a systematic view of the universe for the first time in human history. Then, journeying through time, we will explore the height of ancient science and the magnificent discoveries and inventions that could have sent the ancient world racing forward 2,000 years. Then we will see some of the many reasons that lead to the downfall of ancient science and the coming of the dark ages. Throughout, we will see where much of our world began in words and images, the origin of words and phrases we use every day and come to see that the ancient world was not as far different from today's world as we believe.

Throughout the writing of the book, I attempted to not only tell about the lives of the scientists, philosophers and other notables involved, but also about the world in which they lived. My goal was to also explain (to the best of my abilities) what their governments were like, what they listened to for music and the kind of theatre they enjoyed, along with everyday information: for instance, what kinds of materials were they using for eating utensils? What were they wearing? What sort of decorations did they have in their homes? In addition, there are a myriad of other questions to answer about their world, if we want to understand the people behind the birth of science.

Records from this time are sketchy and there are many conflicting reports. I did my best to gather the facts, compare them with one another and see which conflicted and which pieces fit the general picture. When reports seemed to conflict, I tried to take into account other factors which were occurring in the area, along with general human nature and tried to find the most logical course of events.

Readers of other history books will notice that very rarely do I use the words "According to Plutarch…" or "Pliny states that…" This was done so that the stories would flow smoothly as the reader went along and to try to retain a more coherent, easier style to picture and understand. I hope that I conveyed at least a little of what it was like to live in those times, as far as I was able to research what we know of that era.

One of several problems when writing about history is talking about money from the ancient world and translating it into terms that modern people will understand. There needs to be a standard by which we come to terms with how much money we are talking about when dealing with an ancient unit of currency, such as a

Roman *talent*. What I have chosen to do in this work is to use a gold standard. How much gold would that amount of money have been able to buy then and how much would that gold be worth now? It is not a perfect system, but gold is one of the few good standards we can use to compare the buying power of ancient and modern monetary units.

Starting in 625 BCE, with the birth of the world's first known scientist, Thales, and ending with the final destruction of the Library of Alexandria and the murder of Hypatia in 415 CE, the time period covered by this book encompasses the first period of science, which lasted over a thousand years.

The reasons behind the downfall of this first scientific inquiry beckon to be explored. Many people largely blame slavery for the downfall of science after the first thousand years, but that was only one small part of the reasons for the Dark Ages falling upon the world. The Roman Civil Wars, The Alexandrian War, the lack of an industrial base from which to mass market products, as well as a Christian persecution of Jewish and Pagan people in early fifth century Alexandria all played significant parts in the dramas leading to the Dark Ages. Science remained dormant for a thousand years after that, until it was reborn during the Renaissance.

I owe a special thank you to my father, Robert Maynard, for teaching me science as a child, giving me a rational mind and always believing in me. Thanks goes to the love of my life, Pat LaPree, for putting up with my endless hours of typing at the computer and bringing me coffee, food and hugs. A tremendous thank you to Vesta Hornbeck, professor of English at Keene State College (NH), for editing different versions of this book as it went along, offering advice, correcting errors and reminding me that grammar has changed since I graduated college! Also, thank you to Dr. J. Russell Harkay, also of Keene State College, for encouraging my writing throughout college and trusting me to edit the second and third editions of his book, *Phenomenal Physics*. I would finally like to thank the two authors who filled my childhood days with their words and ideas, as I read their books endlessly: the late Dr. Carl Sagan and Isaac Asimov.

Also, thank you to Ancient Coins Canada Inc. (www.ancientcoins.ca)and Classical Numismatic Group (www.cngcoins.com) for their permission to use several of their images of ancient coins throughout this work. Thanks goes to Liana Cheney of the University of Massachusetts Lowell for giving me permission to use some of her reconstructed music of ancient Greece and Rome on the website for this work (www.lightofalexandria.com).

This book could not have been completed without such wonderful websites from HistoryChannel.com, the University of St. Andrews School of Mathematics and Statistics and Dictionary.com to look up quick facts, check references and teach myself the meanings of words that had escaped my vocabulary for the last thirty-six years.

This book is meant primarily as a work for the curious layperson, but could also be used as a text for science and history courses, as well as for home and private schooling. As the author, I would love to see this work used for all those purposes and more.

Permission is granted to educational instructors to distribute excerpts from this work, in lengths of one chapter or less for educational purposes without any further permission from the author, provided the material is provided free to the student(s). Bulk discounts are available for bookstores, distributors and educational institutions. Please contact me through the website for the book (www.lightofalexandria.com) for any other purposes, including commercial distribution, or for more information.

This work represents four years of work and I sincerely hope the reader enjoys the work I have assembled and learns at least something that enriches their lives.

James Maynard
October 2005
Keene, NH

The Light First Kindled

There is a sort of wonder when we look up into the evening sky. We see perhaps 10,000 stars shine down upon us as we gaze up in amazement.

Nearly every child asks him or herself the same questions: What ARE the stars? How come they do not fall down at our feet? Are there other life forms, circling around those stars, gazing up at our distant Sun and asking themselves the very same questions? And what is this thing called "life"? I remember being a small child and trying to send messages to beings around other planets, using only a flashlight and the few letters of Morse code I learned when I built a telegraph with my father.

The first thoughts of what the stars were and what they were made from, were likely conjured by our earliest ancestors, on the Serengeti Plain, staring up at a multitude of stars, long before the interfering haze of artificial lights.

The stars were also special to our distant ancestors for a different reason. We can touch things on Earth. Rocks, trees, water and even mountains far on the horizon were all knowable. They could be touched, felt and examined. The stars were not that way at all. No matter how high one climbed on a mountain, or up the tallest tree, the stars were no closer. Likely, the early ancients thought that this must be the home of the Gods. Perhaps this quest for the stars was where religion was born, answering the eternal questions with stories, myths and rituals. Most people in the ancient world chose this route, as many people choose it today.

Civilizations around the world attempted to commune, one way or another, with the Gods. Several cultures built temples to the beings in the sky. Many of these were built in an attempt to bring living priests or their honored dead closer to the stars. However, all of these cultures were also subject to the same laws of physics, so the shape of these temples to the sky were all roughly the same. There was only one shape that could be used to build to a great height with ancient materials and techniques: the pyramid. Therefore, this is the reason that we see pyramids from vastly different cultures around the world. There is no mystery to why we see such a commonality between distantly spaced civilizations. There is no need for a super-race of men, or aliens with advanced technology. The quest to become closer to the stars was prevalent worldwide and the only way to get closer to the stars with ancient materials was to build a pyramid.

Mysticism pervaded the world and many cultures stayed in this mindset throughout the ancient years. However, something dramatically changed in the lands of Ionia, in the sixth century BCE (Before Common Era[*]) and a new way of thinking was developed. The area of Ionia was a fertile ground in which the seeds of scientific

---

[*] *BCE (Before the Common Era) and CE (The Common Era) are the same as the more familiar BC and AD. But the BCE/CE format is becoming more accepted today among historians and authors.*

wonder and curiosity were sown.

The people of that distant time and place could see change all around them, as we see it today around us. The movement of the Sun and Moon in the sky, the cycles of night and day, summer and winter, life and death, all beckon to be explored. A brave few of the ancient Ionians and Greeks chose another route. Not of myth and superstition, but of careful observation, study and experiment.

If we lived in a world where change never occurred, science likely would have never been born. There would have been little reason to study nature, for all that would exist would be a static freeze-frame of the world. If, on the other hand, we lived in a world where change happened too quickly, or in such an errant fashion that patterns could not be recognized, science would have died an early death from mass confusion.

However, we live in neither of these worlds. Our universe and the Earth around it, changes according to regular, largely simple patterns and the laws of physics and chemistry are the same under all but the most extreme conditions. We can recognize the patterns of change and predict how materials, mountains and planets will move well in advance.

Science and mathematics are the tools we use to see those patterns, to recognize the underlying patterns and to see the invisible. Science allows us to view the shadows of atoms too tiny to ever make out with the human eye. It allows us to glimpse at ancient quasars, whose light left their source eight billion years before the Earth and Sun were formed. Science allows us to build submersibles to peer onto the ocean floor miles beneath the sea surface where humans would be crushed to death within seconds by the enormous pressures. It also allows us to peek at the surface of Venus, which also has crushing pressures, along with sulfuric acid rain and a surface temperature of 900 degrees Fahrenheit.

Science is unique in human endeavors. Good science works not by people working to prove each other correct, but by scientists trying their best to prove each other wrong. It is as if a team were building a bridge by having each worker not only lay their own struts and jousts, but also hammer away at the struts and jousts laid by others in an attempt to knock it free from its moorings. Many struts will come free using this approach, but the ones that stay will be very secure. This is not done for the sake for spite, or greed or self-aggrandizement. Scientists strike at each other's work for the sake of truth. The best method the human race has ever found for coaxing out the secrets of nature is to allow scientists to test each other's theories, attempting to disprove them. It is only those very few theories that both withstand peer review and which are repeatable that may, one day, stand a chance of becoming known as a true law of nature.

Today, scientists have tested nearly all of Einstein's predictions to better than one part in 10,000. Yet, still, scientists do not refer to the law of relativity. The evidence is not yet good enough for the good scientist to firmly believe, but it is certainly looking like relativity is a very good model for certain conditions and behaviors.

The theory of evolution is in the same conundrum. There is much evidence to back it up. Species on one continent are often very similar (but not the same) to species on a far distant continent if those two continents were once in contact.* Dig deep in the ground and one will see simpler organisms in the older layers (strata) of Earth. However, the greatest proof of the truth of evolution has only come forward in the last forty years, with the discovery of DNA - the blueprint of life.

We have had decades of studying the genetic code of many, many forms of living beings. We can see how species that have similar features have the same strings of amino acids in the same parts of their DNA. We have read the blueprint of life and it shows the road map of evolution. Yet, here as well, no one speaks of a *law of evolution*. Does that mean that scientists today are not sure of the process, as if Darwin had just recently announced the results of his studies of finches he made in the Galapagos while sailing on the HMS Beagle? Creationists often say that evolution has no more precedence in public schools than creationism, because evolution is "only a theory." Perhaps many of the people who make this claim do not understand how scientists use the word *theory*. Almost nothing becomes a law of science. Nearly every model we have of the universe is just a theory. Yet, many of those theories have been shown to be correct to a tremendous degree.

There is a concept in mathematics known as a *proof*. If one makes an assumption that can be formed into an equation and can work both sides of the equation through mathematically legal means producing a statement that no one can argue with, only then can we say that the equation has been proven.

For instance, if I presented you with the wholly unremarkable notion that 6+4 is the same thing as 5+5, if we wished (if just to pander to the obvious) we could say $6+4=5+5$, then solve the equations on each side and produce $10=10$. This is a statement about which few would argue - therefore, our original notion has been proven. It is *only this level of certainty* that scientists will accept as having been proven. *Everything* else is a theory.

A few of the Greek scientists took the bold step of introducing mathematics into their study of nature, securing this vital tool for proving and modeling assumptions. Patterns, once discerned, can be described, predicted and applied to other situations and conditions. It allows us to see the future. Mathematics is the language with which nature speaks.

Just in the last hundred years, science has remarkably changed our lives with airplanes, synthetic materials, fuels and fertilizers. Plastics have changed how we build nearly everything. Electronics have given us radio, television, computers, intercontinental ballistic missiles and instant, inexpensive, personal worldwide

* Not only does this provide good evidence of evolution (since the separated species were once one species and are no more), but it also scores some points for continental drift!

communication. Quantum theory has opened the nature of the Sun's energy source to us and given us microelectronics, nuclear reactors, as well as the ability to destroy our entire planet and all human life, if not all forms of life on the planet, in less than a day.

Yet, as knowledge of science becomes ever more important for both the populous and especially the leaders to grasp, people are falling further and further behind modern research. When important decisions need to be made, such as whether to allow a nuclear plant or a coal burning plant in town, it is far too common for the side with the greatest amount of money, or the best PR people, or the catchiest slogan to win the decision. Meanwhile, science related bills before Congress can swell to mammoth size, include riders that have nothing at all to do with the bill attached to them and the Congressperson has little or no time to read the proposed law or regulation. They will then cast their vote based on faulty assumptions, clever lobbying and political gain.

Science created today's society and lifted us from simple agriculture to the information age. Now there is only one way for people to tend to the dangers and responsibilities of our modern era. That is to understand science.

There were societies and cultures throughout the world which had flashes of insight – notably the Egyptians and the Chinese. Egypt had astronomy that was far superior to anything in the Grecian world at the time of the start of our story and they also had extensive skills in record-keeping and building canals. Nevertheless, as we will see, these talents were mainly put into use for daily living and not for the further exploration of the phenomena uncovered. China also had some great advances in technology well before the rest of the world: inexpensive paper, gunpowder, the printing press, rockets and kites, just to name a few. The numbering system of the Chinese also gave the people of that nation an advantage – unlike the English language today, in the Chinese numbering system, one only needs to learn how to count to ten and all other numbers are derived from those first few numbers. For instance, instead of twenty-two, the Chinese would say "two ten two."

But China had two unique characteristics that kept them from a full flourishing of science. The first was that when designs were made of new machines, the art was considered more important than the functionality of the design. Thus, when new designs were passed on from one person to the other, the look of the drawing was more important than the ability of the person reading it to build the machine, thus creating errors. The other major problem with the growth of science in China was the fact that the people there, since ancient times, considered a career as a government bureaucrat to be the highest calling, largely not considering science a worthy profession.

However, there was a time and place where science first began to flourish: the sixth century BCE in the lands surrounding the Aegean Sea, between modern-day Greece and Turkey. This area is known to us today as Ionia.

Why Ionia?

Ionia was a series of small city-states in what is now western Turkey, in Asia Minor and a chain of small islands off the coast of Greece, in the Aegean Sea. Although they all used the same alphabet and spoke nearly the same language, they employed four different dialects, including one peculiar one which was only spoken on the island of Samos.

A series of invasions by the Dorians, a nomadic people, created a dark age in the Aegean world, which lasted three hundred years, from 1200 to 900 BCE. The areas of Ionia and Asia Minor were first populated by mainland Greeks when they were kicked out of Athens after the fall of monarchy there after the year 1000 BCE. In the years leading up to the Ionian awakening, Mycenae,* which had been a dominant force in the region, began to fall. The Dorians, a people driven from their lands and unwelcome anywhere else in the Aegean world, burned the citadels of Mycenae and conquered their capital city. The people of Greece then began to populate the islands of the Aegean and Asia Minor and took up residence in twelve cities on the western coast of Asia Minor, including Miletus, Clazomenae and Ephesus.

This was the height of classical Greek Paganism as people worshipped Zeus and Apollo, Aphrodite and Athena, among others. They built a great Temple to Hera in Olympia, Greece that stands to this day. Wearing purple became popular throughout the Mediterranean and fabrics were produced which were dyed with alum and purple snails.

The ninth century BCE was the time of Homer, who wrote *The Odyssey* and *The Iliad*, two of the ancient world's greatest works. Both of these books center around the Trojan War which legend tells ended in 1184 BCE, during the dark ages of the Aegean. This war, once thought to be a myth, is now likely to have been based on an historical event. Possibly, the events surrounding the Trojan Horse actually occurred, and the city lost the war due to history's greatest deception. But this was not a war of swords. Bronze bends easily and so swords and daggers were likely a weapon of last resort for soldiers of the bronze age. Most probably, the Trojan War was fought almost exclusively with spears.

When the Aegean woke from its collective sleep, they would emerge like a butterfly from a moth's cocoon, from the Bronze Age into the Iron Age. It seems strange to many people that the Iron Age (in most societies) came after the Bronze Age. However, this had to do with the melting point of the various materials. Ancient kilns could barely melt the copper needed for bronze (90% copper, 10% tin) and fell several hundred degrees short of the 1500+ degrees Celsius (2800 degrees Fahrenheit) required to melt pure iron. Much of this early iron was likely found in the form of a nickel-iron alloy within meteorites. Even stony meteorites (which make up about 90%

---

* Homer refers to this city in the Odyssey and the Iliad, calling her people the *Achaeans*.

of all meteorites) still contain a significant portion of this nickel-iron compound.

Until about 800 BCE, the Greek alphabet (which was based on Semitic-Phoenician letters, to which they added vowels) used only capital letters. After the beginning of the eighth century BCE, lower-case letters were developed in the Greek alphabet. Meanwhile, the Chinese developed their full written script about this same time and the Hebrews developed their own alphabet and began to write literature.

Later, as the classical Greek city-states continued to form, they would soon gather together into the Ionian Confederacy. At the beginning of the eighth century BCE, some Greeks began to settle on the coast of Spain. Later, other Greeks began to settle in Southern Italy, founding Messina and Syracuse, in Sicily; the Spartans founded Taranto, another southern Italian city.

This century also saw the First Messenian War – the first of three conflicts between Sparta and nearby Messenia. The overt cause of this war was the fear of the people of Messenia concerning rapidly expanding Spartan influence and dominance of the area. This revolt leads to the forming of the government and social structure of Sparta into a well-run war machine, under the leadership of Lycurgus. Round one found Sparta the victor, controlling the eastern part of southwestern Greece.

The people of Ionia saw a flourishing of arts and crafts; they began to create pottery with geometric patterns, along with intricately carved griffins (which were used as furniture legs), carefully crafted carpeting and embroidery, along with carvings made from stone. They also produced kouros statues, which were bronze statues of nude youthful athletes, first small, later life-size or larger. Other popular figurines included *kores* (a standing, draped young woman) and small statuettes of seated matronly women. By the time of the birth of Thales, Athens had flooded the Mediterranean markets with vases, which usually had a black background with raised red and white figures sculpted upon them.

The first Olympic games were held in 776 BCE and the author of the day was Hesiod, whose most famous poem, "Works and Days," followed Greek rural life throughout five ages, from a pristine, idyllic past, up to the his time, the Iron Age. This play also reflected the widespread feeling in ancient Greece that the poor should receive little sympathy, unless they also gave of themselves. For in this work, Hesiod states "Give to one who gives, but do not give to one who does not give".[1] That does not mean that charity was unheard of in ancient Greece, but that many people gave freely to their communities, as opposed to individuals.

Approximately twenty years after Hesiod wrote *Works and Days*, a new city state would form in Italy, which would later come to have a great effect on Ionia and the rest of the world as Rome was founded upon seven hills in the western central arm of the Italian Peninsula. The people of this city would go on to form the greatest republic the world had ever seen and then fall into an often despotic dictatorship as the future people of this empire-to-be began to rule the Mediterranean world.

---

[1] Finley, M.I. "The Ancient Economy." Berkeley: University of California Press. 1973.

There were other powers already on the Italian Peninsula that would challenge Rome in the early years of the city. The most powerful and dominant of these were the Etruscans. Originally either from Lydia or perhaps Italian natives, this society was ritualistic and exercised great control over her people. Even family life was dominated by an all-powerful government. They would also extended their control over much of the Italian mainland throughout the sixth century BCE.

An example of the Etruscan style toga which became a Roman favorite.

Here it is shown worn draped around the body in the older style.

Public Domain Photo

The Etruscans were a people acclaimed for their gifts of prophecy and known for being great builders; it was from the Etruscans that the Romans learned to build aqueducts. They were also the first civilization of which we know that constructed marble statues and clay figures of people.

A Roman woman wearing a tunic. This style was popular throughout the Ancient Mediterranean.

Public domain photo

The role of women in Etruscan culture was a dichotomy. In many ways, they were expected to be very subservient, not only to their husbands, but to any man who desired an evening with her. During their drinking parties, couples would pair up for lovemaking, often within sight of each other. In addition, every Etruscan woman was expected to prostitute herself once in her lifetime for an offering at a temple.

However, the Etruscan women were also given freedoms that even the Greeks and Romans would never extend to their women. The Etruscan women were allowed to go to public events and sit with the men; this was unheard of in the ancient world. More importantly, Etruscan women kept their names after marriage and the names of their children would reference their mother and perhaps their father. Unlike all of the other ancient Mediterranean civilizations, the Etruscan households were guided largely by the woman of the house, who had great discretion in how to raise her children and in her style of dress.

Many of the Greeks and Romans considered these traditions barbaric, but the question needs to be asked about how much of this was due to actually moral reprehension. After all, the Romans were famous for their public adult activities after a dinner and night of drinking, as well. It is possible that much of this apprehension was due to their distaste of the public and home life of the Etruscan women.

It was from the Etruscan alphabet that the Romans would develop their own alphabet, which (with a few modifications) we use today in English. Their language, however, was totally unlike all the other languages in Europe. It appears to have developed entirely independently of any outside influence.

Etruscan art (used strictly for practical and religious purposes) also greatly influenced later Roman art works, as well.

One practical Etruscan development used later in Rome was their style of helmet for the military. This helmet allowed easy visibility, while still providing a good deal of protection for the soldier wearing it. Eventually, the Roman military uniform would be a mixture of the best armor found among other civilizations, but the "Roman" helmet seen so often in movies and pictures was taken directly from the Etruscans.

An Etruscan helmet from the seventh century BCE. It was this style helmet that the Romans would adapt into their own uniforms.

Photo believed to be public domain.

Another adaptation from the Etruscans by the Romans was the method of dress; The Etruscan style toga became a Roman favorite, when people wore them, which was not as often as many might believe. Most citizens of Greece and Rome wore a tunic and the ones who did wear a toga, usually wore them over a tunic.

At first, Rome was a monarchy, but the post of king was not an inherited position, as it is in most cultures. For the early Romans elected their kings, who served until they died. Many Romans were brave, but they were also a superstitious people, believing in the power of omens and living their lives at least partly to please the Gods, who they saw all around them.

Early Rome was made up largely of escaped slaves, bandits and mercenaries. This early influence produced a population that would not lie passive in the face of tyranny or aggression. When the threat came from the outside, they became the finest fighting force in the world, carrying symbols of their mascot, the wolf. When the threat of tyranny came from within, they were the most dangerous subjects any leader could imagine and no leader was safe.

The Romans had a tradition in declaring war that existed (in a sense) even into the late Republic, in the first century BCE. War could not be declared solely by the king, or the later consuls. They would bring their grievances first to the potential enemy themselves and make an offer that would avert war. If the land in question did not accept these terms, the proposal of a declaration of war would go before a panel of high priests, known as the *Fetiales*. Once this board determined that the war was just, a member of their group would head to the border of the land in question and throw a spear into the land of the enemy. Only then was war officially declared. As the Republic grew larger, it became difficult to bring these members of the Fetiales to the actual border in question. Therefore, when war was to be declared, the Senate would dedicate a portion of the Senate room to the country or city-state to be attacked and the spear would be thrown into that.

Rugged survivalists, the Ionians were a strong and determined people. The area they chose to live in was marginal for farming and they quickly found they could grow little but olives and a few grapes for wine. In fact, the nature of growing olives played a significant part in the development of Mediterranean culture and social organization, for the olive tree is not a labor-intensive crop, but it does require a great deal of patience, stability and an investment in long-term care, as it does not bear fruit for the first ten or twelve years. Therefore, the people of the area could not live a nomadic lifestyle; once the trees were planted, the farmers needed to tend their crops for over a decade in order to reap the benefits of its fruitful bounty.

For in addition to olives as food, the fruit was also the most important source of soap and fuel for the people of this time and place. Olive trees also survive through severe summer droughts, thereby not requiring the type of mass collective construction projects required in Egypt for irrigation farming. Thus, people gathered in smaller groups, with more of an emphasis on local politics and community. Whereas the irrigation-farming societies suffered greatly when the central bureaucracy inevitably fell apart from time to time, causing mass starvation and pestilence, the smaller, decentralized communities of the Aegean could quickly recover from natural and manmade disasters.

A typical scene of everyday life in Ancient Greece. The woman plays a lyre and the scene shows a *kores*-like statuette.

Public Domain photo

Turning to the sea for survival (not to mention a break from eating olives and grapes), they found themselves surrounded by two great maritime powers: the Babylonians to the east and the Egyptians to the west.

Science never took root in either Babylon, nor in Egypt. The climates there were regular and the people believed that all they would ever need to know was knowledge needed for the day-to-day life of an agricultural society. If they knew when to plant which crops and how to take care of them, then they had all the information needed for life. The

Egyptians had a strange flirtation with science, but again, it was mainly directed towards the practical problems of everyday life, such as answering the question "When is the Nile going to flood again?" The greater questions of life were answered, for them, by a great belief in the super natural and a rich and varied afterlife.

One of the first coins in the world: an electrum Lydian coin from about 600 BCE, with a lion pictured on the front of the coin.

Photo courtesy Classical Numismatic Group

This was not the case in Ionia. There, the marginal land, hostile neighbors and the fact they had no room for expansion made these people begin to think differently. The lack of any official state religion made the people of Ionia reject the idea of theocratic kings and instead, they governed their city-states by consent of the slave owners. The slaves in ancient Ionia were debtors, who were slaves to their creditor until the debt was paid off and then they were released.

In the sixth and seventh centuries BCE, the Lydians, who lived in what is now northwest Turkey, ruled the whole region. They were known at that time for the splendor of their magnificent capital, Sardis. In their greatest contribution to our own time, they first introduced coins as a method of exchange for goods and services. The coins were made of electrum (60% gold and 40% silver) and were traded throughout Ionia and Greece. The Lydians also gave us the word *tyrant*, which, in the language of Lydia, meant *lord*. Their government was a series of oligarchies: rule by the few.

By this time, the Ionians had established great trading routes dealing in copper, gum, grains and salt, among other products. To aid their trade, they produced the first maps we know of today.

A mosaic showing a male Roman tibia player and a woman dancing and playing the *crotala*, pairs of rattle-like instruments.

Photo believed to be public domain

This same era also saw an explosive growth in epic poetry and song, sung to the accompaniment of lyre and a type of flute known as a tibia. Tibiae (the plural of tibia) were often made of wood and later, bone (sometimes even human bone). Players who were proficient with this instrument would sometimes even play two of them at once, providing their own harmony. This instrument would remain popular, even through the Roman Empire.*

* There is a limited amount of reconstructed Greek and Roman music, including the sound of tibiae,

When life would end for these Greeks and Ionians, the body was carefully washed and anointed with oils. The deceased was then wrapped in two layers of cloth and a gathering was held. At the ceremony, people would sing music and play instruments, later bringing the body to a cemetery, where they would be buried in the ground and memorialized through either a stele (similar to a tombstone) or, if they could afford it, a small bust upon the grave.

By 545 BCE, a Persian defeat of the Lydians left the area under the control of a distant empire (roughly modern Iran), barely able to keep an eye on their distant quarry. This encouraged Ionian society to an even greater level of respect and desire for diversity and competition. There were over two hundred major religions in Ionia, but a few individuals chose to believe none at all.

The alphabet of the Ionians was the most important of the Eastern Greek alphabets and the version used in Miletus was later made the official alphabet of Athens in 403 BCE. It was partly the development of this rich, complex way of communicating that made the Ionian language the first one to ever be both complex and flexible enough to allow scientific discussion of any depth.

Even with the first light of science, the seeds of its destruction were already being sown. For although Greece and Ionia would offer the world great minds, theories and discoveries, there was little innovation for the common person. Most of this early scientific advance was considered "off limits" for those who were not philosophers. This created a paradigm in the minds of the Ionian people that there were certain castes of intellect which those who wished to learn should never cross. The advances in science and technology were never made available to the common person, who saw little reason to believe or trust the scientists.

There was little experiment in the ancient world, but experiment started at nearly the same time as true scientific theory. A man named Anaximander was the first experimental scientist and he was best of friends with the first theoretician, Thales of Miletus.

---

courtesy of Liana Cheney of the University of Massachusetts Lowell, at www.lightofalexandria.com.

Thales and Anaximander 625 - 546 BCE

Thales of Miletus

Photo believed to be public domain

Born in the city of Miletus, in what is now the country of Turkey, in 625 BCE, Thales name would become synonymous with knowledge and thought. Ancient Greeks and Ionians might refer to an especially bright (or foolish) person "a Thales," much as we today might refer to someone as "an Einstein." He would become known as one of the seven sages (wisest people) in ancient Greece.

Thales was the first person we know of to reject superstition and declare that everything could, one day, be understood in terms of science, logic and reason. Thales was a theoretician, attempting to unlock the secrets of the universe with pure thought, imagination and a scientifically voyeuristic mind.

He appears to have been nearly entirely self-taught. Except for studying with the astronomer/priests of Egypt for a while, there is no record of anyone having Thales for a student. While in Egypt, he may even measured the pyramids, by measuring the shadow they cast on the desert sand. He simply noted the length of the pyramid's shadow at the time when the lengths of the shadows of people were equal to their height. At that time, he knew, the length of the shadow of a pyramid would also be equal to its height.

Anaximander of Miletus

Photo believed to be public domain

Thales seems to have spent some time in the army and when his unit was caught near a river that they could not cross, Thales had an idea. He directed the digging of a trench behind the army, in the shape of a crescent, beginning and ending at the river. Sure enough, much of the water diverted through this trench, limiting the flow of the river in front of the army, allowing them to cross.

After possibly engaging in trade during his days in Egypt, he led a modest lifestyle. One winter, people were chiding Thales for not working more to enrich himself monetarily. Thales had seen signs that the olive crops the following year were going to be abundant. Therefore, he raised what capital he could and bought the use of all the olive presses in Miletus and Chios for the following year. This came cheaply, as no one even thought to bid against him. When the winter ended and an abundant olive crop did arise, he rented out the use of the presses at an inflated rate and became wealthy. He explained to people afterwards that he wanted to prove that a philosopher could get rich if they chose to, it is just that time spent on business would get in the way of studying nature.

He is regarded as the first person to ever do a systemic study of astronomy and is credited with first recognizing the constellation of Ursa Minor (The Little Bear). He was also the first person to carefully study the path that the Sun and the Moon make through the sky and became the first person to be able to predict eclipses.

One day, it is said, Thales was walking outside with a servant-girl named Theodorus, who was known for her beauty and wit. He was explaining his theories about the stars to her and likely looking towards the sky when he fell into a well. Theodorus, laughing at him, stated that Thales wished to know about the objects in the sky, but that the things right under his feet escaped his notice.

Anaximander was ten to fifteen years younger than Thales, possibly born in 611 BCE and some historians state that he was a pupil of the elder philosopher. He was also the first person to attempt to make a detailed analysis of all of nature. Unlike Thales, Anaximander was a careful experimentalist and inventor. In fact, it is very likely you may have one of Anaximander's inventions in your home or garden, for Anaximander was also the inventor of the sundial.

He was also light-hearted, and Anaximander was not afraid to laugh at himself. Once, he was told that children would laugh when he would sing. Anaximander replied "We must then sing better, for the sake of the children."[*]

Just as the world's first two scientists should, they often disagreed. Thales believed that the world and everything in it, developed from water and that the Earth itself rested upon an infinite sea. Earthquakes, he believed, were the shaking of the Earth from waves within this cosmic ocean. Anaximander rejected the idea of an infinite sea and proposed instead that the Earth was free-floating, unsupported by anything. He thought that perhaps the whole universe was bowl-shaped, with the Earth resting in the middle.

Thales envisioned a process, much like the forming of a delta at the mouth of a river, which he believed was responsible for the formation of land above the waters. Although his idea of the formation of the land is not correct, his method of answering the question would help bring about modern science. It wasn't the right answer, but it was a reasonable assumption, based on a logical train of thought: Thales knew water can carry sand and silt and that the sand and silt can collect forming dry land. Perhaps that is where all dry land originates, Thales reasoned.

Anaximander reasoned differently; he believed that all things we see in the world around us sprung from an "indefinite" which permeated all space and time. He thought that the motion of this indefinite caused the opposites to separate from one another, producing hot and cold, wet and dry, etc. This predated the later idea of the

---

[*] Diogenes Laertius. "Lives of Eminent Philosophers.". *Anaximander*.

eternal permanence of matter.

The world around these two men was quickly changing. The Areopagus,* a council of elders, which acted as a court and who named three Archons to lead the city, led the unpopular government in Athens. In 632 BCE, just seven years before the birth of Thales, Cylon, a former star of the Olympic games, led an unsuccessful coup in Athens, supported by his father-in-law Theagenes' naval fleet. The uprising was completely put down and Cylon escaped the city with his brother, never to return.

The Assyrian Empire, which, earlier that century, had destroyed Babylon and diverted the Euphrates River (now within the borders of Iraq) to flood what remained of the city and who had sacked the cities of Memphis and Thebes, began to fall. The Chaldean general, Nabopolassar, took Babylon and became their king, declaring Babylon free and independent of Assyria. Finally, Medes, Babylonia and the Scythians defeated Assyria at Nineveh, finally ending the Assyrian Empire, which was divided amongst her conquerors. Throughout the Aegean world, the worship of Apollo and Dionysus (also known as Bacchus) became more popular.

The Athenian archon Draco was pressured to write down and codify the laws of Athens which previously had only been an oral set of rules. He therefore enacted the first written laws in Athens in 621 BCE, when Thales was a toddler and a few years before the birth of Anaximander. They were cruel, harsh laws, where even small crimes, including stealing a head of cabbage, were punishable by death. It is from these severe laws of Draco that we get the term *draconian laws*, meaning laws where the punishment is far more severe than is appropriate for the transgression. Draco claimed that severe punishments were the only means of deterrence for small crimes and that for larger crimes, he could not think of a harsher punishment than death, but that if he could, he would use it.

Then, in 594 BCE, the great lawmaker and statesman Solon came to power. Athens had been ravaged by an agricultural depression, which had left many of her farmers in the chains of slavery and her people in need of the food the farmers had produced. He immediately cancelled all debts and mortgages on bodies (agreements where slavery was the result of defaulting on a loan) and encouraged those who could no longer participate in agriculture to take up other professions, fueling crafts and trade. He invited foreign businessmen into the city, reformed the currency and coinage and brought the poorest people into the political realm by re-structuring the caste system into four groups, based upon the land and property owned by an individual. This extended to the lowest class the right for them to participate in the public assembly, although they were still forbidden to run for political office.

However, this caste system caused a unique kink in the Athenian economy.

---

* The Areopagus derived its name from the low, rocky area west of the Acropolis, where it convened. This is the same place that the Bible says St. Paul preached his Sermon to the Athenians and where, in Greek mythology, the Goddess Athena supposedly reigned over the trial and acquittal, of Orestes for the murder of his mother, Clytemnestra.

Citizenry was extended only to those males who owned land and were not currently serving out a debt of slavery with a creditor and ownership of land, with very few exceptions, was limited only to citizens. Therefore, the legal citizens of Athens made up only about 15-30% of the population of the city. Much of the rest of the population was engaged in successful trade, manufacturing and lending money. However, the non-citizens who lent out money that would have fueled the economy found that they could not take land as collateral, because they had no power of foreclosure. Because of this, large loans were difficult to obtain and to lend, which stifled growth for the rest of classical Greek history, even through complete changes in governmental systems.

In addition, Athens in the time of Solon ran a great trade deficit. Athens, for the most part, exported only some olives and olive oil, along with wine, silver, potting clay and marble. However, keep in mind that these were natural resources, which were common throughout the region. Therefore, supply and demand kept the prices for these export goods low. Meanwhile, Athens was importing nearly two thirds of her grain, as well as their entire quantity of needed ship timer, iron, copper, tin, flax for linen and papyrus. The city was barely self-sustaining in meat, fish and wool. Money kept flowing out of the city faster than it was coming back in to her merchants and citizenry. The effect of this is an inevitable economic downturn and lower buying power for the people of the city.

Solon, who was also a poet, also formed a new body, the *boule*, a body of 400 members, who were elected from the aristocracy and whose duty it was to prepare legislation for the general assembly, or *Ecclesia*.* The power to prepare these bills had been previously held by the *Areopagus*. The Areopagus held real power in Athens, as they named the Archons (leaders) of the city. In return for having stripped them of the power to introduce bills to the assembly, Solon gave the Areopagus the power to try officials and Athenian citizens of crimes of morality: crimes against the perceived well-being of her citizenry.

Like most philosophers in the centuries to come after him, Thales spoke both of natural phenomenon and human nature. In these long-gone days, the two fields were seen as nearly one and the same and anyone who spoke of natural science was expected to also offer discourse as to the nature of human thought, behavior and emotion.

Thales, for instance, stated that one should never enrich oneself by shameful means and that a person should expect to receive as much help from their children as they had given to their own parents when they were younger.

Even in modern times, scientists such as Carl Sagan and Stephen Hawking

* In 508 BCE, the Athenian statesman Cleisthenes increased the size of the boule from 400 to 500 members, increased the power of the body and extended the right to serve on the boule to all citizens in good standing. This, of course, meant you could still not serve if you happened to be a woman, a non-property owner, or paying off a debt through servitude.

continue this tradition to an extent and find themselves making comments in popular books of science about nuclear proliferation, population growth and government funding for super-collider experiments.

There is a story that Solon went to meet with Thales and enquired to the philosopher as to why he had never married, or had any children. Thales secretly had an actor friend of his come into the building later and pretend to have just arrived from Athens. When Solon asked him of news from the city, the actor told the Archon about a funeral for a young man, whose name he could not recall, but who was the son of a wise and powerful Athenian, who had been gone from there for quite some time. The actor led Solon on and as Solon became increasingly concerned that the funeral in question was for his own son, he began to break down and hit his head in frustration. Thales came over to the grieving Solon and confessed the deception. He stated that if terrible events like that could cause even as great and strong of a man as Solon to break down, then Thales wanted no part of family life.

Supposedly, in his youth, his mother had begged him to marry and Thales kept replying to her, "It is not yet time." When he grew older and his mother asked him again, his answer was that, "It is no longer time." When people inquired to him why he did not have children, his standard reply was that it was because he was "fond of children."[1]

A bust of Sappho of Lesbos; one of the greatest poets of the ancient world.

Public domain photo

Also in the time of Thales and Anaximander, on the island of Lesbos (or Lesvos), a woman named Sappho led a group of young maidens in a school of poetry. It is said that her poetry was filled with emotion and was beautiful and simplistic, but much of what which exists today are mere bits and pieces of the many works she wrote. Only one complete poem remains today, *Ode to Aphrodite*, reproduced on the next page.

Sappho was born on Lesbos, in Mitylene, around the year 610 BCE, the only daughter out of three children. Having been born about that time, she was approximately the same age as Anaximander. Her mother Kleis had also been born on the island and Sappho's daughter Cleis would also be born on the island.

Small and dark-skinned, she wrote at least nine volumes of poetry and developed what became known as the Sapphic verse form of poetry. In this style, the four lines consist of eleven syllables each, except the fourth line, which contains just five syllables.

As each of her students left the school to become married, Sappho would

[1] Diogenes Laertius. "Lives of Eminent Philosophers." *Thales*.

compose their wedding vows, as a farewell gesture to those leaving her domain.

But for you, O Dika, bind your hair with lovely crowns,
Tying stems of anise together in your soft hands.
For the blessed graces prefer to look on one who wears flowers
And turn away from those without a crown.[1]

It is said that Sappho herself, who also played the lyre, was in love with another poet of the time, Alcaeus. She may have even jumped off a cliff to her death at age sixty over an unrequited love for Phaon, a young sailor.

During her life, she is also said to have had several female lovers, to whom she addressed much of her poetry. They all lived with her at her school and the most well known of these women were named Erinna of Telos, Damophyla of Pamphylia and Atthis.

Perhaps partly inspired by the works of Homer over two hundred years earlier, this was a period of great, epic poetry and stories. Besides Sappho, the Aegean was filled with many other poets and lyrists, including Kallinos, Archilochus, Tyrtaeus, Mimnermus and Alcaeus.

Elsewhere in the world, Zoroaster (in Persia) and Lao-Tse (in China; the founder of Taoism) reached prominence. Construction began on The Marduk Temple in Babylon, the infamous Tower of Babel.

Ode to Aphrodite

Many colored throned immortal Aphrodita,
daughter of Zeus, wile-weaver, I beg you
with reproaches and harms do not beat down
O Lady, my soul.

But come here, if ever at another time
My voice hearing, from afar
You gave ear and your father's home leaving
----golden --- you came
yoking the chariot.

And fair, swift doves brought you
over the black earth
dense wings whirring, from heaven down
through middle air.

Suddenly they arrived and you, O Blessed One,
Smiling with your immortal countenance
Asked what hurt me and for what
now I cried out

And what do I want to happen most
In my crazy heart. "Whom then Persuasion
.............to bring to you, dearest? Who
Sappho hurts you?

And if she flees, soon will she follow,
And if she does not take gifts, she will give,
If she does not love, she will love
despite herself"

Come to me now, the harsh worry
let loose, what my heart wants to be
Done, do it! and you yourself be
my battle-ally.

English translation by William Harris
Text in public domain

In Greece, temples began to be constructed with marble and limestone and

[1] From Anne Carson, "If Not, Winter – Fragments of Sappho." 2002.

houses became more detailed and intricate. Life-size sculptures of women became popular and the first Ionic columns were raised on the Island of Samos. Pottery began to move from geometric patterns to images of animals and people, as the Greek city-states became more influenced through trade with the Phoenicians.

The seven-string lyre was invented and music consisted of singing, accompanied by that instrument and tibia. One composer (and poet) at the time was Arion, who introduced the concept of the strophe and antistrophe, terms for parts of a choral arrangement associated with Greek dramas.

Thales offered insight into geometry, being the first to define the diameter of a circle, as well as recognizing the mathematical symmetry of the isosceles triangle. He was also the person who first developed the notion of a "rigorous mathematical proof" that was discussed in the first chapter. This helped Greek mathematics evolve from simple arithmetic to a study of mathematics and geometry. Beauty in mathematics was seen for the first time as the subject came one step closer to the modern notion of mathematics: the study of patterns.

Thales is known for having predicted the eclipse of 28 May, 585 BCE, which halted a battle between the Lydians, under King Alyattes and the Medes, led by King Cyaxares. Despite the fact that Thales had warned the Lydians beforehand that the eclipse would take place, both sides quickly broke off battle, fearing it a sign of foreboding from the Gods.

Anaximander had important ideas about astronomy and evolution. He taught that all life originally sprung from water and that all water was permeated by "daemons", close to our modern knowledge of bacteria. He even went further, saying that humans had evolved from fish by adaptation to the environment. These were among the earliest ideas on the subject of evolution.

In addition, Anaximander produced what may have been the first map ever of the known world, which stretched out to Persia in the east. Herodotus, who saw some of these early maps, wrote that they were round, divided in half at Delphi, with the Mediterranean in the middle and the surrounding land encompassed by the River Ocean. The maps labeled just the lands surrounding the Mediterranean as the habitable world and the north was said to be populated by mythical northern people and the south by the black "burnt people."

Thales had established a school in Ionia and the last student there was Anaximenes, another resident of Miletus. Anaximenes may have also studied under Anaximander, as well. He had been born in the year 570 BCE and was in his mid-twenties when Thales died. Anaximenes was the first person to talk about rarefaction* and condensation of air and taught that air would become visible when cooled and condensed. This was a reasonable assumption for someone who lived thousands of years before anyone would see liquid nitrogen or oxygen.

It is possible that Thales may have been seriously hurt in the fall down the well,

---

* Decrease in pressure

for Anaximenes wrote to Pythagoras about the "death" of Thales, saying "while he was looking up towards the skies, he fell down a precipitous place. So now, the astronomer of Miletus has met with this end... But we who were his pupils cherish the recollection of the man and so do our children and our own pupils: and we will lecture on his principles. At all events, the beginning of all wisdom ought to be attributed to Thales."* Apparently, the reports of his death were greatly exaggerated.

Near the end of the life of Anaximander, Croesus was conquered by Cyrus the Great of Persia and the cities of Greece came under the control of the Medes, near Persia. The people of Ionia began to prepare for war, in the cause of universal human freedom and out of fear of slavery. However, just at this moment, Persia attacked the Medes and defeated them, deposing their king, Astyages, the son of Cyaxares. However, because the Medes had fought so successfully in the century leading up to that war, the Persians considered the people of Media to be equals, the two civilizations united under the Persian banner and now Ionia and Greece were under Persian control.

Ionia and Greece were now ruled by the recently expanded Persian Empire, as the first two scientists met their ends. The younger Anaximander died first, in 547 BCE and the elder Thales passed away the summer of the following year, when he became overheated while a spectator at a gymnastics event. Meanwhile, a twenty-two year old with a fondness for travel was about to take an old trick of farmers and realize the importance of a simple mathematical relationship.

---

* Diogenes Laertius. "Lives of Eminent Philosophers." *Anaximenes.*

Pythagoras 569 - 475 BCE

If one wishes to be a math or science student, s/he will be inundated with the so-called "Pythagorean theorem". The idea is first presented in junior high perhaps, but the idea will surface again in studies of trigonometry, first and third year calculus and geometry.

For those who may not remember the formula, it simply states that in any right triangle (a triangle having one corner of 90 degrees), the sum of the squares of the lengths of two sides is exactly equal to the square of the length of the hypotenuse (the side opposite from the right angle). This allows the calculation of the length of any one side of a right triangle if you know the lengths of the other two sides. $A^2+B^2=C^2$.

Pythagoras of Samos

Photo believed to be public domain

Although Pythagoras gets the credit for the formula that bears his name, the idea itself may have been known in ancient Greece for two thousand years before his birth. Although little is known of the thought processes at that point in history, we do know that the ancient plows these people used could not easily turn. To make up for this problem, farmers would tie a rope with twelve evenly spaced knots and create a triangle in the famous 3-4-5 relationship. This allowed them to create three square farm plots along the sides of the triangle, through which a farmer could drive his plow straight through from one plot to another. Pythagoras (or one of his students) discovered that this idea applied to all right-angled triangles. It is told that when he discovered this relationship, he sacrificed one hundred oxen to the Gods in celebration.

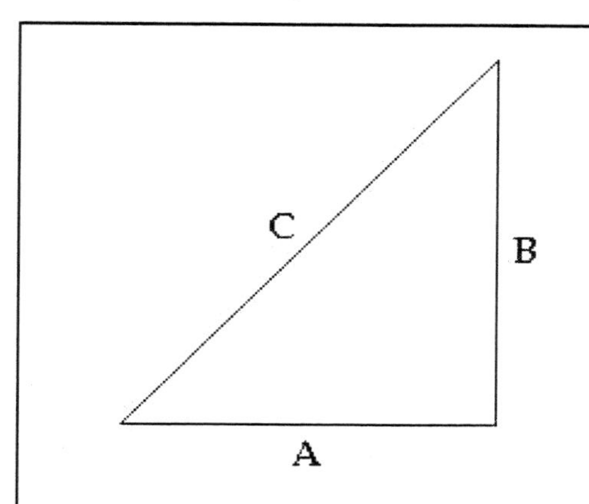

Pythagorian Theorem:

The length of the hypotenuse (C) in a right angle triangle is equal to the square root of the sum of the squares of the lengths of the other two sides (A^2 + B^2).

Pythagoras was from the island of Samos, where he was born in the year 569 BCE. He was the son of Mnesarchus, who was a merchant and likely a Phoenician. His mother, Pythias, was a native of Samos. Mnesarchus was granted citizenship in Samos for bringing a load of grain to the city during a famine.

As a child, Pythagoras traveled widely with his father, to Syria (where it is said he learned from the masters there) and Italy. Even as an adult, Pythagoras continued to travel to many lands, including Egypt and the great cities of the

Babylonian Empire (which encompassed part of what is now western Iraq).

Around the time of the birth of Pythagoras, the second-to-last of Rome's kings, Servius Tullius, determined that he needed to how many potential soldiers there were within Rome. Thus, he decided to have them counted and assigned people to do so. The people were gathered into several groups, based on wealth. Since the Roman soldiers were expected to provide their own arms when joining the military, this would also give Servius an idea of how many weapons and horses could also be brought to bear in time of war. Those people who were too poor to provide their own weapons were exempted from military service and were counted just by their heads (*capite* in Latin). This part of the count of the population then became known as the *Capite Censi*, or head count. It is from this that we derive our modern English word "census." This first census in history determined that there were around eighty thousand men in the city of Rome who were able to provide and bear arms. After Tullius, Tarquin the Proud, the last king of Rome, feared that an idle population was a dangerous population and set the people counted in the Capite Censi to work on mammoth public works projects.

The greatest mathematician up to his day, Pythagoras talked of the geometry of all triangles. He may also have been the first person to recognize the five regular solids. These are three-dimensional objects, whose faces are all regular polygons. Four of these were discovered fairly quickly, but the fifth, the dodecahedron (made of 12 pentagons), was discovered later and was considered sacred. The news of its existence was barred from public knowledge for many years.

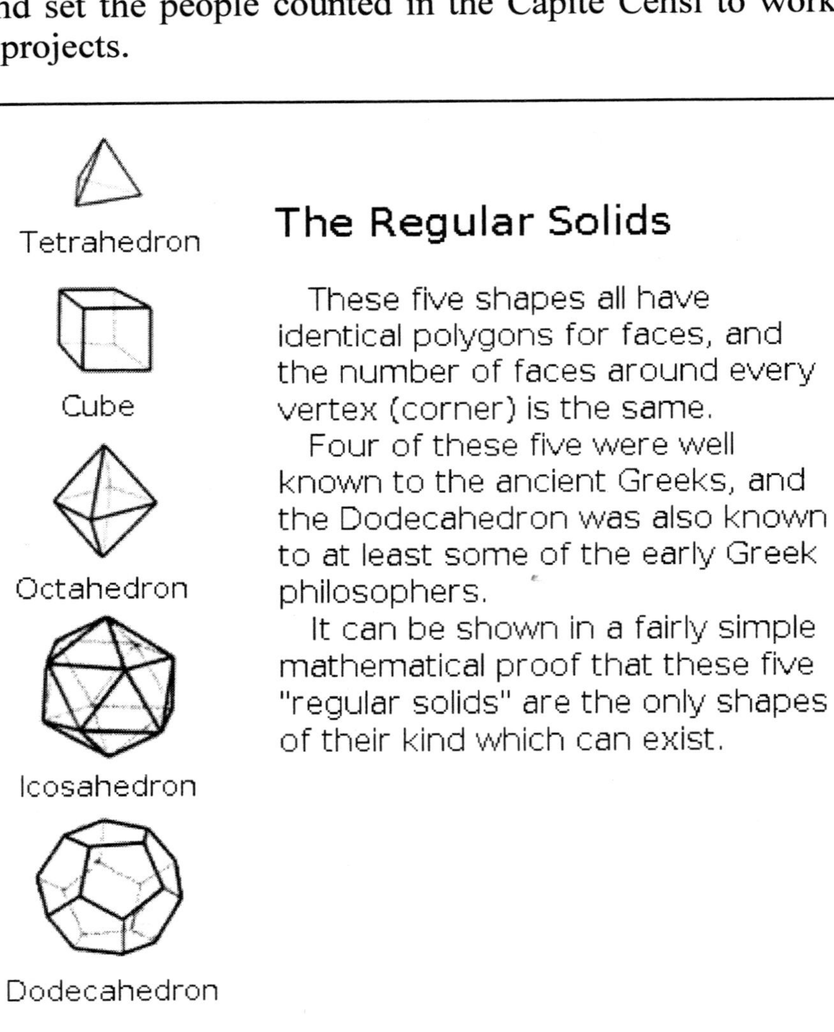

The land Pythagoras occupied was divided - caught between reason and superstition, myth and science and fear and understanding. Because of this, little was written on the life and death of Pythagoras and what was written tends to portray him almost as a God, all wise and knowing.

An Ionian hemiobol, circa 510 BCE. This is one of the smallest coins ever minted.

Courtesy Ancient Coins Canada Inc.

He had been a student in Egypt in his youth, where he learned the Egyptian language. He also adopted their methods of teaching and later taught his students using the Egyptian methods, for which he became widely ridiculed. This became known as the Italian school and would be adopted by several philosophers for their students in the coming years.

For five years, his students were not allowed to speak or even see the master and for seven, they could not question their distinguished teacher. The thought was that after seven years of intently listening to the great master without question or objection, they would come to see the wisdom and truth of his words. The students, whose numbers exceeded 600 for every lecture, also kept all their goods together during this time, in a sort of communal colony.

His school was also highly mystical and his students may have greeted each other with secret symbols and Pythagoras gave advice to his followers in hidden metaphors, such as "Do not stir the fire with a sword", meaning "Do not anger and engage the wrath of powerful men".[1]

They renounced worldly pleasures, practiced self-denial, being careful not to eat, drink or engage in love to excess. They also ate a vegetarian diet, except for beans, which were considered impure and that he believed caused nightmares[*]. He also usually refrained from eating any food that had been cooked, preferring a raw diet instead. He refused to drink wine during the daytime and he would usually eat vegetables for desert, either raw or occasionally boiled. In this polytheistic world, Pythagoras would only worship Apollo, for the offerings given to Apollo did not include any meat and only raw foods were offered to this God of healing and prophesy.

Strangely enough, however, he may have recommended a diet high in meats to athletes, who until that time trained on a diet of cheese, figs and breads. Then again, this recommendation may have come from another person at that time and area, an

[1] Diogenes Laertius. "Lives of Eminent Philosophers." *Pythagoras.*

[*] Although it is possible that his advice to "Abstain from beans" may have had nothing at all to do with eating after all. Black and white beans were used as method of voting in Magna Graecia (southeast Italy) and it may have been advice to his followers to stay out of politics. Pythagoras himself, however, seems to have engaged heavily in politics.

athletic trainer who was also named Pythagoras.

He was always well dressed in bright white, very clean clothing. He also kept his woolen bed sheets very clean and bright white. Pythagoras was never known to engage in small talk, or even to tell a joke or humorous anecdote. He considered such small talk to be beneath the dignity of an educated man. Even when he was angry with someone, he would never become visibly upset with him or her, even if that person were a slave.

After leaving Egypt, he landed in Crete. Here, he entered the most holy spot for the people of Crete, the cave of Idaea. This was not the first time he had entered a holy area, as he is reported to have done the same while he was in Egypt. He wished to know the religions and philosophies of the areas that he traveled and was not afraid to traipse upon solemn ground in order to do so.

He left Crete to return to Samos, which was now under the rule of Polycrates the tyrant. However, the opposition of many people to his school's policies (which included admitting women as students) may have, at least in part, led to him fleeing Samos for Crotona, in southern Italy, around 518 BCE. There are also reports that he was highly successful in influencing the region's politics and that a rival party may have encouraged his ouster.

Like Anaximander, he too believed that the Earth was free-floating, unsupported by anything. He also realized that the Moon's orbit is inclined to the equator of the Earth by about five degrees. Pythagoras was also the first to recognize that the "morning star" and the "evening star" were, in fact, both the same object; Venus. He even told that the Moon shone by reflected light of the Sun.

Pythagoras believed that the lower atmosphere was constantly in motion, in an age when people were still arguing over the existence of air as a physical material. Furthermore, he taught that the air near the ground was permeated by disease and that these disease-causing beings were both alive and mortal, foreshadowing knowledge of germs. Well ahead of his time, Pythagoras realized that both plant and animal life had to spring from other life and could not form from inanimate material.*

However, Pythagoras always had a strange fascination with the perfection of the heavens. Differing from his theories about the lower atmosphere, Pythagoras taught that the upper atmosphere was always at rest and pristine. The beings in the upper atmosphere, he believed, were immortal and divine. The Sun, Moon and stars, he believed to be Gods, due to the life-giving warmth of the Sun.

Nevertheless, mathematics and pure science were not all that fascinated Pythagoras. For he also was the first person we know of to do a systemic study of music. In doing this, he recognized the mathematics inherent in musical notes.

---

* This is not quite true. All life is based on DNA and the building blocks of this master code are amino acids, which can be formed from inanimate material. If you give amino acids enough time and enough energy in the right environment, they can form simple life and evolve over billions of years into complex organisms.

For instance, if we take a string held tautly at both ends like a guitar or piano wire and vibrate it, it will produce a note. A similar string with similar tension that is half the length will produce a note one octave higher. This is because the number of vibrations produced per second would have doubled. One example is that if our first string produced 440 Hz.*, known as middle A, our second string would produce a note of 880 Hz., also an A, but one octave higher. Raising a note by 50%, say from 400 to 660 Hz., is what we call a fifth in music. The note produced at 660 Hz. is an E note.

Pythagoras was also the first person in Greece to establish a standard system of weights and measures. This odd philosopher/scientist may have also been the first person to ever scientifically investigate the sport of boxing and develop techniques that he attempted to pass onto youth by boxing with them. When he became ridiculed for boxing with the youth, he donned a purple robe and boxed adult men, winning several matches with his newly developed methods.

Nearly thirty years younger than Pythagoras, another philosopher/scientist was developing his theories on the universe about this same time. This was Heraclitus, who was born in Ephesus, a large, rich Greek city in Ionia.

Heraclitus was an arrogant man, yet he was aloof, preferring quiet contemplation to the banter of social gatherings. This behavior was interpreted by many as a sign of hatred towards people in general and it may have held some truth. He would speak of theology, yet likely angered many people by his ridicule of mainstream religion and ceremonies. Due to these aspects of his personality, he became known as *the weeping philosopher*, or *the dark philosopher*. When he was asked once why he remained so taciturn, Heraclitus replied, "That you may talk."[1]

Politically, Heraclitus tended towards authoritarianism, even once having proposed the idea of executing all the youth of Ephesus and banishing the children from the city in retaliation for the exile of his friend Hermodorus from the city. Perhaps fortunately for the people of Ephesus, when Heraclitus was asked to make laws for the city, he refused due to his belief that the constitution of the city was too flawed.

Hercalitus may have studied under Xenophanes, but he rejected teachers for the most part and was largely self-taught. He believed that if one understood a problem, then all truth would fall from the model. His model of the Earth was based on fire, believing this to be the prime element from which all else sprung and in which everything was finally dissolved.

There is only one book that is known to have been written by Heraclitus, entitled *On Nature*. This work is divided into three sections, covering the universe, politics and theology. Unlike what is told of the usually clear writing style of

---

* Hz. Stands for Hertz and is simply another way of saying cycles per second. It is named after the German physicist Heinrich Rudolf Hertz (1857-94).

1 Diogenes Laertius. "Lives of Eminent Philosophers." *Heraclitus*.

Heraclitus, *On Nature* was written in a purposely-obscure style. This was likely done in order that the well-learned people might understand it, while it would remain safe from the criticism of the common people. The effect of this was that after the death of Heraclitus a group of people calling themselves the Heracliteans was formed, a sect who centered their philosophy on this book.

Heraclitus spoke of the evaporation of materials from both earth and sea, some of which was visible and some not visible. Heraclitus believed that these evaporations were responsible for weather and the cycles of day and night, as well as the seasons.

The ideas of Heraclitus on astronomy also partly concerned the Sun and the stars. He was correct in his belief that the Sun is a nearby star and the stars were distant suns. However, he also believed that the Sun and stars were physical vessels like large pots of fire gathering the evaporations. Eclipses, he reasoned, were seen at times when the bottom of the vessels was turned toward the Earth. The phases of the Moon he attributed to the turning of the vessel holding the Moon as it circled our home world.

Lucretia; the woman whose death inspired the overthrow of the monarchy in Rome.

Photo believed to be public domain

The last of the Roman kings, Tarquin the Proud, ruled from 534 to 509 BCE. He would lose not only his seat as king, but also the monarchial system itself over a crime committed by his son, Sextus, in 509 BCE.

Sextus had raped Lucretia, the wife of one of his friends, Lucius Tarquinius Collatinus. After she told the story to her husband and father, she pulled out a knife and stabbed herself, committing suicide.

The people of Rome were outraged by this act on the part of the son of their king and they rose in revolt against Tarquin. The leader of the rebel forces was Lucius Junius Brutus, a distant ancestor of the Brutus who would assassinate Caesar over 450 years later.

Rome became a republic and the era of kings there was ended. At this time, their control stretched for only 24 km (15 miles) measured from north to south and less than 50 km (30 miles) from west to east.

The Palatine Hill, where Rome was first founded, became one of the richest areas in the city and was home to some of the wealthiest people in Rome. It is from the word *Palatine* that we get our modern word *palace*.

During his life, Pythagoras wrote several books, on politics, education, history and natural philosophy. A student named Lysis of Tarentum later compiled many of the books and poems of Pythagoras into collected works of the great master.

He had a wife named Theano, who was a student of his and a resident of Crotona. She bore him a daughter named Damo and a son named Telauges. In 479

BCE, four years before the death of both Pythagoras and Heraclitus in 475, Persian control of Greece ended after the Battle of Plataea.

Heraclitus spent his last days retired at the Temple of Diana with his children. While here, he became an avid dice player. One day, some onlookers berated him for taking up the habit of gambling. Heraclitus, evidently still upset over what he perceived to be the undesirable government of his home city replied, "Is it not better to do this, than to meddle with public affairs in your company?"[1]

The diet of Heraclitus in his last days consisted mainly of wild grass and plants and his body soon started filling with fluid; his medical condition would today be known as *edema*. In order to drain the fluids from his body, he holed himself up in a stable and covered himself with dung, believing that the warmth from the decaying fecal matter would dry up the fluids inside his body. He died soon after, at age seventy and his body may have been eaten by dogs.

An epigram written after the death of Heraclitus read, "I've often wondered much at Heraclitus, that he should chose to live so miserably and die by such a miserable fate."[1]

Pythagoras may have been murdered by a snubbed dinner guest in 475 BCE while at the house of Milo, a famous Olympian athlete who dressed in a lion-skin cloak and carried a club. The guest was considered by many at the dinner to be unworthy of dining with the assembled group and he went outside and set the house on fire, possibly with the assistance of some of the townspeople of Crotona. As the victims ran outside, nearly all of them were slaughtered. Either Pythagoras was killed here or he escaped but starved to death while in hiding. Lysis was at this dinner, but was one of the few survivors.

The works of Pythagoras were entrusted to his daughter Damo, who rigidly held on to them, forsaking a potential fortune, because she knew that her father would not have wanted to see the works sold. Telauges, the son of Pythagoras, took over the school upon his father's death and he may have taught Empedocles. Within fifteen years, his school became divided and more political. Soon thereafter, rival forces wiped out the people who remained at the school. Today, nothing remains of the works of Pythagoras. And much of what we do know, we do not know whether to credit to Pythagoras himself, or to a student such as Lysis, due to his demand for an uncritical student body who quoted the master long before expressing their own ideas.

---

[1] Diogenes Laertius. "Lives of Eminent Philosophers." *Heraclitus*.

Anaxagoras 500 - 428 BCE

Anaxagoras was born into a wealthy family near the year 500 BCE, in the city of Clazomenae, one of the original twelve Ionian cities of Asia Minor. At twenty years old, he left for Athens, where he stayed for thirty years, studying everything he could in this rich environment. He was also a student of Anaximenes, possibly visiting him in Miletus.

Anaxagoras of Clazomenae

Photo believed to be public domain

Athens was becoming the center of the Greek world at this time and Anaxagoras thrived in the bustling atmosphere. In return, Anaxagoras brought the rational study of nature from Ionia to Athens, where it would reach its next great peak. In only one hundred years, Athens would in turn lose its place as the center of science due to war and the rise of a tyrannical government.

Anaxagoras certainly seems to have had a following while here in Athens. Euripides, the famous epic poet of the day, developed an appreciation for science through the influence of Anaxagoras. He may very well of even taught Socrates, when the famous philosopher was still a student.

As an adult, he chose to renounce his wealth (giving it to his family) and instead, to devote himself entirely to science and philosophy, which he considered to have been the reason for his birth. Throughout his life, he also steered himself entirely clear of any involvement with politics.

His family berated him for not taking care of his part of the estate. "Why then," he answered, "do you not take care of it?" When people would state that he had no affection for his country because of his disdain of politics, he stated, "Be silent. For I have the greatest affection for my country." as he pointed to the sky.[1]

He was a scientist, philosopher and artist. He was a man who believed firmly in a mechanical universe, much in line with the view of the universe that was held in later years, by Kepler, Newton and Galileo.

Anaxagoras began to use geometry in the study of astronomy and very nearly developed the correct explanation for the causes of eclipses. Strangely, he believed that in addition to the Earth causing shadows upon the Moon during a lunar eclipse, there were additional dark bodies between the Earth and Moon that also cast their own shadows. The reason for these additional bodies is unclear from what is known of the work of Anaxagoras.

---

[1] Diogenes Laertius. "Lives of Eminent Philosophers." *Anaxagoras.*

It has also been said that Anaxagoras designed stage scenery for plays painted so that objects that were to be seen as being further away were painted smaller and closer objects larger. This may not seem like a remarkable insight to our modern worldview, but it was a major advancement in art, that would not be rediscovered until the beginning of the Renaissance.*

For his great insight into nature, he was given the nickname "Mind", by the people of Athens:

"They say too that wise Anaxagoras, deserves immortal fame; they call him Mind..."[1]

As people praised the work of Anaxagoras, a woman named Aglaonike took astronomical observation and a keen mind to a new level. She expanded the work of Thales and learned to predict both lunar and solar eclipses with such precision that people were astounded. However, because of her gender, people attributed her predictions to sorcery and not to science. She was the first female astronomer we know of and she is almost entirely forgotten today.

When Anaxagoras was in his early forties, Aeschylus wrote *Oresteia,* a trilogy of plays about a family blood feud. These plays went on to have a great influence on writers in the coming decades and centuries. In 449 BCE, Herodotus releases his work, *History*, telling of the events of the past, including the Persian Wars, which had lasted from 490 to 479 BCE, ending in an uneasy alliance between Athens and Sparta.

The year after the end of the Persian Wars, Athens embarked on a massive build-up of her navy, conquering city-states along the Aegean coast. Many of these cities grew tired of the constant fighting and attempted to withdraw from Athenian control, but they were kept in check through the use of the mighty Athenian naval forces. In 464 BCE, there was a massive earthquake near Sparta that left many people dead. In the chaos and confusion which resulted, the helots in Messenia revolted, and Sparta called upon Athens for assistance in 462. However, there had recently been a political power-grab in Athens, and when the Spartans heard of this, they immediately sent the Athenian forces away, refusing to fight alongside them. This broke the fragile alliance and Sparta and Athens were once more at odds.

Anaxagoras began his cosmology by declaring that the Sun was not a God (but was, instead, a red hot stone) and that the Moon, which he stated contained hills and valleys, shone by reflected light from the Sun. He even postulated that the size of the Sun was larger than Peloponnesus: the peninsula in southern Greece containing Sparta

---

* When perspective was rediscovered by Leonardo Da Vinci, Brunelleschi and Piero della Francesca, this was one of the greatest events leading to the full flowering of the Renaissance. More on that in the second book of this series.

[1] Timon, Greek philosopher and poet (320-230 BCE). From Diogenes Laertius. "Lives of Eminent Philosophers." *Anaxagoras.*

and Corinth. This was considered a tremendous size, in the days when the Sun was thought by most people to be no larger than a shield. He also foretold the idea of orbits, stating that the objects in the sky stayed together by circling quickly around one another and that if this velocity slowed, the circling body would come crashing back to the object that it was circling.

His theory about the Sun was reasonable and the last notion is a very good explanation of the causes of orbits. The notion of reflected sunlight from the Moon also shows the remarkable insight of this man. For, by speaking of the formation of the planets from the same ball of dust and debris that formed the Sun, he also showed a remarkable knowledge of centrifugal force and rotational inertia.

Anaxagoras theorized that comets were planets breaking free from their places in the sky (releasing sparks that we see as shooting stars) and letting out light as they traveled. He also believed that the Milky Way was lit by reflected light from the Sun.

None of these ideas is currently accepted, but he had some other ideas about science that were far ahead of his time. In an age when most people believed the stars were eternal, fixed and unchanging, Anaxagoras taught of stars moving in the sky over the course of eons, as well as the changing of large geological features given enough time. When asked whether the mountains of Lampsacus would ever become sea, Anaxagoras replied, "Yes, if time lasts long enough."[2]

He also worked in optics, formulating theories about light and rainbows. In meteorology, he taught that thunder was the sound of clouds crashing into one another and lightning was formed by clouds rubbing together. His notion of lightning was fairly close to the truth, as lightning can be caused by positively and negatively charged clouds passing too close to one another.

The Acropolis in Athens today. Two people are standing on the steps on the right side of this picture, under the second and third columns from the right, giving a sense of scale.

One dry day, Anaxagoras showed up in the town of Olympia, wearing a leather cloak, something that was only normally worn in rainy weather. The people around him may have been amused at his choice of outer gear, until it actually began to rain that day.

When Anaxagoras was reaching his 50s, a new statesman rose in Athens. His name was Pericles and he and Anaxagoras became friends. The ability of Anaxagoras to draw those who would make history around him would now spell trouble for the astronomer.

In 447 BCE, Pericles began

[2] Diogenes Laertius. "Lives of Eminent Philosophers." *Anaxagoras.*

construction on the Acropolis in Athens, which would go on to host some of the most famous events in Athens in the coming centuries. It was also in this period that the Athenian Empire reached its pinnacle. For ten years, starting in about 440 BCE, Athenian control extended along most coastal regions of the Aegean. Athens was now the major super-power in the region and the city would soon find itself with the burdens of her newly-found responsibilities. For instance, the Athenian court system, which had always accepted oral testimony as the most reliable form of witness, began to rely more heavily on written testimony and evidence.

The insistence on written testimony in court, along with the great deal of inscriptions that have been found from Ancient Greece suggest that most people of that time had at least a functional literacy. Opponents to that idea (including William Harris in his 1989 work "Ancient Literacy") dispute that idea based on the fact that Greece did not (for the most part) have a state-sponsored educational system. However, many of today's studies show that home-schooled students in the United States perform at least as well as their public-school counterparts, and so it seems that home-schooling in Greece would have easily been able to produce a literate population.

Those who opposed Pericles began to oppose Anaximander as well and this would lead to his downfall. The citizens of Athens passed a new law (likely introduced by opponents of Pericles) that it would be a crime to not practice religion, or to teach theories about objects in the heavens. Anaxagoras was arrested in the year 450 BCE (approx.) for his teachings about the heavens and possibly on charges of sympathizing with Persia. This was also a crime in many other Greek city-states of the era.

While in prison, Anaxagoras began studying a problem in mathematics known as *squaring the circle*. This is simply a way of finding a square with the same area as a given circle, using only a ruler and compass. Why was he attempting to solve this with a ruler and compass? We will hear several times of people in Ancient Greece attempting to solve mathematical problems using just these tools. That is because there was a belief at this time that ALL mathematical problems should be able to be solved using these two tools: a ruler for straight lines and a compass for curves.

Mathematics in Ancient Greece also had another couple of peculiarities. For one, they had no different set of characters in their language(s) for numbers, so they used letters. Alpha was one, beta was two, gamma was three, etc. Another difference to our modern numbering system was the fact that they had no decimals in their method of counting. Every number less than one had to be written as a fraction of two different numbers. This is the primary reason that when irrational numbers (such as the square root of two, which can not be expressed as the ratio of two numbers) were discovered, there was widespread disdain of this idea. It is as if we were told today that some mathematicians had discovered a new number that could not be expressed in decimals; many people would dismiss it and most people would ignore it.

There is a similar modern concept of complex or imaginary numbers. These actually do exist and create real effects in the physical world, but cannot be found on a "one-dimensional" number line. Most people believe that imaginary numbers are just that; but they are not.

Around the year 450 BCE, the people of Rome decided that they needed to codify their laws, as had been done in Athens under Solon. With an unwritten body of laws, aristocrats reigned supreme since the consuls decided not only the interpretation, but also the nature of the laws. Since no laws were written down, no one had a basis for appeal of a conviction. Therefore, a group of three prominent Romans left for that city in 450 BCE, returning three years later. Their recommendations were carried out and for one year, Rome had no consuls, but was instead ruled by ten men in an executive body known as the *decemviri*. This board both ran the day-to-day operations of the city, along with preparing a written set of laws to be voted upon by the people the following year. One of the members of the *decemviri* was Gaius Julius, a distant ancestor of Julius Caesar.

Pericles had his friend freed from prison, but Anaxagoras was forced to leave Athens and headed to the Greek city of Lampsacus, where he founded a school. Lampsacus was a center for the worship of fertility and procreation and it was here that he died in 428 BCE. After his death, his birthday became a holiday for children in Lampsacus that was celebrated for centuries after his death.

Empedocles 493 - 433 BCE

A woodcut of Empedocles

Public domain photo

Empedocles was a physician, philosopher, politician and poet born in the city of Acragas, which is now Agrigento, Sicily, in the year 493 BCE. His father's name was Meton and his name, Empedocles, came from his grandfather, a wealthy man who raised racehorses.*

Agrigento (or Agrigentum) was a prosperous city of about 800,000 people. When he grew to be an adult, Empedocles perceived a strange dichotomy in the lives of the people of that large city. He stated, "The men of Agrigentum devote themselves wholly to luxury as if they were to die tomorrow, but they furnish their houses as if they were to live forever."[1]

When Empedocles was thirteen, in 480 BCE, the Persians sacked and nearly destroyed Athens, in the first of the Persian Wars. First, he learned under Parmenides, along with Zeno. They left the school together, but Zeno went on to found his own school, while Empedocles opted for further study, learning from Anaxagoras, who taught him a great respect for physical experiment.

Empedocles may have attended lessons held under Pythagoras when he was a teenager, but then he was kicked out of the lectures for reprinting the lessons of Pythagoras in poems. A law was then passed forbidding any poet from attending the lectures of Pythagoras.

Ever a proponent of democracy in Athens, he became the leader of the Democrats in Acragas and after helping to overthrow the cadre of rulers in Agrigentum, the people there offered him their crown as king. He refused the offer and instead helped to form a democracy in the city. He also paid for the dowries of the women of Agrigento who could not otherwise afford to marry. Later, after the political situation had changed, he and his followers fled into exile in Peloponnesus.

He was wealthy, arrogant and pretentious. He wore his hair long and dressed in the robes of kings, with a gold bracelet and slippers with soles made of bronze. Empedocles was nearly always accompanied by a retinue of young boys to attend to the great master. He claimed that the knowledge that he possessed gave him royal privileges and that he should therefore be treated as royalty. He seems to have developed this regal bearing while learning under Pythagoras, imitating his styles of life and dress.

---

* Other ancient sources, including Telauges, the son of Pythagoras, dispute this lineage, although most other sources still have Empedocles growing up around racehorses.

[1] Diogenes Laertius. "Lives of Eminent Philosophers." *Empedocles.*

Empedocles even claimed to have been able to perform miracles. He may have, at one time, kept the corpse of a dead woman from decomposing for a month. After this event, talked about by several contemporaries, he declared himself a prophet. Empedocles certainly did not suffer from a lack of confidence.

He was a great orator and skilled in the use of metaphors to make a point. He was an early developer of the method of debate known as rhetoric and wrote dozens of plays in his youth, including several well-liked tragedies.

A poem by Empedocles, telling of his skills and abilities:

And all the drugs which can relieve disease,
Or soften the approach of age, shall be
Revealed to your inquiries; I do know them,
And I to you alone will them disclose.
You shall restrain the fierce unbridled winds,
Which, rushing o'er the earth, bow down the corn,
And crush the farmer's hopes. And when you will,
You shall recall them back to sweep the land:
Then you shall learn to dry the rainy clouds,
And bid warm summer cheer the heart of men.
Again at your behest, the drought shall yield
To wholesome show'rs : when you give the word
Hell shall restore its dead.

From Diogenes Laertius. Lives of Eminent Philosophers. Empedocles.

Text in public domain

When furious winds began to damage the crops where he lived, he ordered that livestock be slaughtered and their skins made into large bags. He was able to determine where to set up these bags, so that they caught much of the wind, saving the crops and the people. For this act, he was greatly admired, receiving the name "wind-stayer" in tribute.

Empedocles taught that human body had four humors: blood, bile, black bile and phlegm. This notion went on to dominate medical thinking until after the time of Shakespeare.

He believed that everything we see around us was composed of earth, air, fire and water. Although we now know that none of these materials is a true element, it was his belief in the simplicity and beauty of nature that still inspires scientists today.

He also taught that the air was a real substance, not just a void, as many people of the time believed. He did this by using a common household device at the time, known as a *water thief*. This was used for the same purposes we would use a ladle for today. A water thief is metal and has a sphere at the bottom and a long neck coming out from the top. There is a hole at the top of the neck on which you could put your thumb and several holes in the bottom of the sphere. People would simply immerse the sphere in water and then put their thumb over the hole on the top to capture the water. Something,

awful explanation

Empedocles reasoned, must be holding that water in place. He realized the only reasonable answer was that air existed and that it could exert enough pressure to overcome even the force of gravity. Using just a simple kitchen utensil and the power of his mind, he had uncovered the workings of an invisible material.

Empedocles wrote on the interaction between harmony and discourse between particles, making an early mention of what became the study of entropy.

He spoke of the Sun being larger than the Moon and he believed that light traveled at a finite speed. After all, he reasoned, if light had to travel between the stars and the Earth, then certainly it took time to travel, thus light had to have a finite speed. Like many of the Greek and Alexandrian scientists, he once again came to the correct conclusion centuries ahead of his time.[*]

He also believed that the only kind of change possible was movement through space. In fact, there *is no way* of defining change without referencing a movement through space. If you measure time by a clock, you are referring to the movement of the hands, or changes in the tags in old-fashioned digital clocks and which pixels or LEDs are lit up in modern digital clocks. Heartbeats are movements of one's heart muscles through space. Water clocks, the positions of the Moon and Sun, the movement of a pendulum and radioactive decay all measure change as movement through space[†].

When the city of Selinus on the southwest coast of Sicily was caught in a plague causing death and miscarriages, Empedocles noticed that the plague was accompanied by foul smells rising from the Selinus River. He ordered the diverting of two other rivers into the Selinus. This was paid for at the expense of Empedocles and the plague lifted soon after. The people of Selinus held a festival soon afterwards, at which Empedocles appeared as a guest. The festivalgoers lifted him up in praise and they prayed to him as they would a God.

He was also a song composer, who even wrote his treatise in verse, being the last of the Greek philosophers to write in such a style. His songs were sung at the Olympics in 440 BCE. He was also said to have a vivid imagination, an eloquent speaking manner and a flair for the theatrical.

Empedocles was at a dinner one day and the slaves and servants who were serving the dinner were serving copious amounts of wine and no food. Empedocles ordered them to bring food, but the dinner host stated that the guest of honor had not arrived yet. This may seem strange to us today, yet dinner was a very formal event in

---

[*] This fact would not be confirmed until the early nineteenth century by measuring the exact observed timings of Jupiter's four large moons.

[†] Incidentally, this shows *a deep connection* between space and time. For time has to be referenced to as a type of change, whether it is the movement of clock hands, radioactive decay, etc. And in order to have any type of change, there has to be a movement through space. Therefore, time can only be measured as the change of position of one or more particles or objects through space. In 1905, Einstein would unite the concepts of space and time, but for different reasons.

ancient Greece and guests at dinner parties would usually sit on stools, except for the guest of honor, who would occupy the sole chair present at most tables.

When the guest did make his appearance, he ordered all the guests to either drink wine, or have it poured over their heads. Empedocles held his words at that moment, but the following day he had both the guest and the host executed, fearing the tyrannical nature of anyone who would give such an order. This was just at the beginning of his political career.

There is even a legend that tells of a legendary death, although it is unlikely to be true. Supposedly, his aura of invincibility led him to claim that he was immortal. To prove this remarkable claim, Empedocles is said to have thrown himself into the mouth of Mt. Etna, located on Sicily. However, the more likely story is less glamorous, as he may have died at a feast in Peloponnesus, or by falling off of a chariot, breaking his leg, which later became infected. After his death in 433 BCE, none of his students felt worthy to carry on his work.

Empedocles believed that he had remarkable knowledge and abilities, but at the time of his death, there was a twenty-seven year old philosopher who developed theories two thousand years ahead of his time. His name was Democritus.

Democritus 460 - 370 BCE

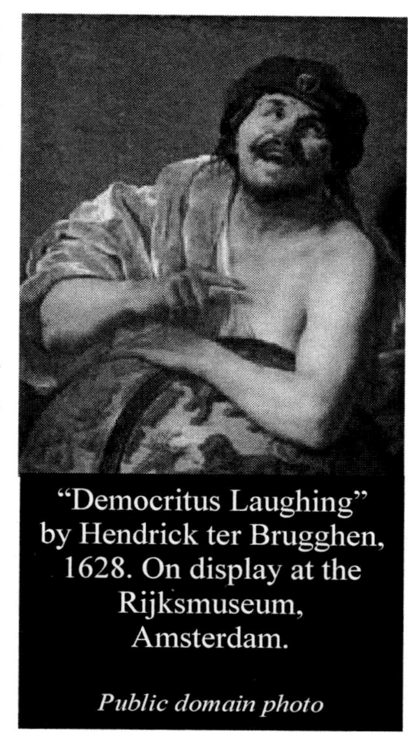

"Democritus Laughing" by Hendrick ter Brugghen, 1628. On display at the Rijksmuseum, Amsterdam.

Public domain photo

It is estimated that during his lifetime, Democritus of Abdera wrote seventy-three books, none of which survive to our present day. It is reported that in these books, Democritus stated that the Sun was just a star and the stars were all distant suns. He believed that a retinue of planets accompanied many of these stars. He spoke of cold, dark worlds, devoid of a stellar companion and how these worlds, including the Earth, are formed and later destroyed.

He was born in 460 BCE, on the island of Abdera, in the same year that Hippocrates, the father of modern medicine, was born on the island of Cos. Abdera, in its time, became known for its fortification walls, dockyards and harbor but it was a laughingstock of the Hellenistic world. Jokes were even told about people from Abdera, portraying its people as slow and dim-witted. Democritus would prove to be a worthy exception.

Settlers from two of the original twelve Ionian cities, Clazomenae (in the seventh century BCE) and Teos (in 545 BCE), founded the city of Abdera, far in the northern Aegean. In Abdera, Xerxes, known as Ahasuerus in the Bible, led his famous army from 486-465 BCE. His army brought Egypt under Persian control once more and brought Athens under its total submission.

Also from Abdera was Leucippus, who was born ten years after Democritus, yet became his teacher. Leucippus believed that the stars were lit on fire by their motion through the sky and that the stars set the Sun on fire. Nevertheless, he was somewhat correct that the reason that lunar eclipses are much more common than solar eclipses is due to the Moon being far closer to the Earth than the Sun. Leucippus first began to ponder systematically on the interaction between atoms and how they would react with one another. Democritus would later expand on that theory, sensing knowledge 2,300 years ahead of his time.

Perhaps it was the sea and the maritime life around Democritus that made him begin to journey through thoughts on the processes of life. Although Anaximander had spoken earlier of evolution, Democritus was the first person to speak and write at length of evolution. He believed that all life developed from simpler life and that humans too, evolved from simpler beings. Democritus, 2,300 years before Darwin and Wallace, spoke of life initially arising within a "primordial ooze."

Democritus was a world traveler in the days long before easy travel. He visited Egypt, Persia, Aethiopia and India. When his father passed away, Democritus and his two brothers split the family estate into three parts. Democritus took the money, about 100 talents (around one million dollars today), the smallest of the three shares, since he needed the money for travel. He spent his entire inheritance traveling the world.

He was the first to postulate that the sensation of taste had nothing to do with the inherent nature of the atoms, but that the size and arrangement of the atom was responsible for this sensation (more or less true). Democritus also put forth the idea that the position of the atoms (which he was clear to differentiate from shape) created what we perceive as color. In this theory, a white object has smooth, flat atoms that cast little to no shadow. Black materials, he pictured as rough, casting many larger shadows. This theory is incorrect, but it was a brilliant idea for the time and was an early truly scientific hypothesis.

Over 2000 years before Newton and Leibniz, Democritus began to discover calculus. He and his teacher Leucippus would work on the *method of exhaustion*, which allows one to measure areas indirectly, by producing closer and closer estimates of the true answer. Democritus also talked of how the volume of any solid object (say, an apple) could be calculated as the sum of the surface areas of a great many planes (in this case, circular slices of the apple). The idea of using numerous planes to calculate a volume is just one small step away from the discovery of integral calculus.

Democritus also worked on the idea of conic sections. The concept of a conic section is such that if one were to take a cone and slice it parallel to the base, the cut section would be in the shape of a circle. If the slice occurred at an angle, the created shape would be an ellipse and so forth. He also realized that in making such a slice, the circle nearer the point would have to be slightly smaller than the other face of the slice, for if it were not, than the original shape would have had to be a cylinder and not a cone.

Democritus was also the person who first coined the word *atom*, meaning *indivisible* and he first developed the modern notion of an element (a substance with only one kind of atom; for example, oxygen, or hydrogen, or uranium). Today, each of the ninety-three naturally occurring elements have been accounted for and studied. He taught that all that exists are atoms and the void between them. All else, he believed, was merely thought to exist. The atoms moving randomly through space, he believed, would crash into one another attracting more material. This was an early notion of the idea of accretion, which is today believed to be responsible for the formation of the solar system.

Democritus was not the first to speak of atoms; his teacher spoke of them, as well as Anaxagoras, but he went forward another step. Democritus began to classify elements into groups. Today's periodic chart (not based on Democritus' classifications) is not set up in any random fashion. Elements within groups in vertical columns on the periodic chart act very much like one another chemically.

However, Democritus also spoke poorly of Anaxagoras, stating that the ideas of

Anaxagoras were "stolen" from older sources. Of course, in his youth, Democritus may have attempted to have become a student of Anaxagoras and was rejected, fueling a dislike of Anaxagoras on the part of Democritus.

When Democritus traveled to Athens, he was already well known and Democritus despised the fame that his teachings had imparted to him. He built a small room in his garden, where he would go hide away from those who wished to be near the philosopher. He would go and listen to Socrates, yet he never introduced himself to the other philosopher, who was ten years his elder. Surprised and relived, he stated, "I came to Athens and no one knew me."[1]

He taught that belief in gods rested in a desire to explain natural phenomena and his school of philosophy taught that the ultimate good is cheerfulness. He wrote several books on ethics, two of which are entitled "Of Virtue" and "Of Tranquility."

In 438 BCE, the Greek architect Ictinus and his assistant, Callicrates, completed work on the Parthenon. This was built on the ruins of two earlier temples, the old Temple of Athena and old Parthenon, which had been destroyed by the Persians in 480 BCE, before work was completed on the project. The structure was located on the Acropolis in Athens, the raised, fortified area of the city.

Just a few years later, in 432 BCE, Sparta and Athens went to war in the Peloponnesian War. For all that is heard today about Sparta, it was never a large or grandiose city. They never had more than 9,000 inhabitants at any point in time in Sparta and the buildings were largely small family dwellings, not large public works projects. The people of Sparta also never embraced physical science, although they loved history and archaeology.

One interesting aspect of Sparta was their relationship to a group of people called Helots. The Helots were a group of people who worked the estates and lands of Sparta and who are often referred to as slaves. However, they were unlike slaves in the modern sense: they could not be bought or sold and they could (and did) raise their own families. However, they also could not be released from service except by the state and they could not go off to find their own land. In addition, unlike the typical slaves in Ancient Greece, the Helots were drafted into the regular Spartan army, in combat duty. Most of the other slaves who were used by the Spartans were in support roles as cooks, carrying supplies or setting up and tearing down camp.

At the beginning of the fourth century BCE, the city-states of Athens and Eretria rose in revolt against the tyrannical rule of Persia under Darius I. The revolt was quickly defeated and Persia set out to punish the people of the two cities, which triggered the Persian Wars.

Democritus considered himself the most traveled and best-informed philosopher of his time and likely with good reason. He had spent all of his fortune in travel and returned penniless. A law in Abdera ruled that no one who had withered away his inheritance would be buried in his home city. Not wanting to have the fate of his body

---

[1] Diogenes Laertius. "Lives of Eminent Philosophers." *Democritus*.

left to chance after death, Democritus read his collection of greatest works (The *Great Diacosmos*) aloud to an assembled crowd, who erupted in cheers. They not only gave Democritus 500 talents ($5 million in today's dollars), but also erected bronze statues to him, elevating him to nearly a deity.*

Democritus died about 370 BCE at around ninety years old. The last few days of his life were the summer festival of Thesmophoria, a three-day event for women only, to celebrate Demeter and Persephone. The sister of Democritus was concerned that the philosopher would die during this holiday and prevent her from making the required sacrifices. Democritus told her to keep her spirits up and go to the celebration. For three days, he had loaves of freshly baked bread delivered to his house and kept himself alive by breathing in the warm steam emanating from the loaves. He passed away painlessly the following day.

After his death, Plato wished to burn all the works of Democritus, fearing the lessons inside of his books. But two other philosophers, Amyclas and Clinias, prevented him from doing so, stating that the works of Democritus were already so wide-spread that burning the copies they could gather would do no good in ending his theories.

The secrets to the universe uncovered by Democritus and his remarkable insight were astounding. He wrote on a variety of subjects, including the books "On the Planets," "Of Nature," "On the Nature of Man," "Causes of Celestial Phenomena," "On Geometry," "On Poetry," "On Rhythms and Harmony" and many more. Today, there is a crater on the Moon named after this brilliant man and just northeast of Athens sits the Democritus Nuclear Research Laboratory.

---

* It is also possible that the family of Democritus read this work aloud after the death of Democritus and the crowd gave them 100 talents, so that Democritus could be buried in his hometown.

Hippocrates 460 - 370 BCE

Hippocrates of Cos; the father of modern medicine.

Public domain photo

Being born (and dying!) in the same years as Democritus, but on the island of Cos, Hippocrates became the father of modern medicine. The Hippocratic oath is named after him, although some aspects of the oath were likely developed by Hippocrates himself, while other parts were added later on by other physicians. However, they were the same ideals (including "Above all, do no harm," maintaining professional conduct and the right to privacy for a patient) under which he practiced.

Hippocrates wrote many books, chronicling his vast knowledge for future generations on a variety of subjects related to medicine. Through his research and writing, the influence of Hippocrates would be felt for centuries. In one book, entitled *The Oath*, he wrote the idea that doctors to this day still make a prime belief:

"Into whatever houses I enter, I will go into them for the benefit of the sick and will abstain from every voluntary act of mischief and corruption; and, further from the seduction of females or males, of freemen and slaves.

Whatever, in connection with my professional practice or not, in connection with it, I see or hear, in the life of men, which ought not to be spoken of abroad, I will not divulge, as reckoning that all such should be kept secret."[1]

History tells that his father was also a physician, but medicine itself would change radically once the Hippocratic (or Coan) school became established. Here, for the first time in human history, medicine was treated as a scientific subject and the rules for good science, including a critical mind and objective reasoning, would become the tools of medicine.

It is likely that Hippocrates learned under the tutelage of Herodicus, a physician who practiced and taught in Thrace. Herodicus is considered the father of sports medicine, promoting the use of low impact exercise and massage for muscles recovering from wounds.

Hippocrates may have also studied in Athens, but returned home to Cos. It may have been while in Athens that the ideas of Pythagoras, Democritus and Empedocles influenced his thinking. This was also about the same time (c. 400 BCE) that the first baby rattle was invented by Archytas of Tarentum. Aristotle wrote that such a device was useful to keep children amused so that they did not break things around the house.

---

[1] Hippocrates. "The Oath." Circa 400 BCE. Quoted in the MIT Internet Classics Archive. <http://classics.mit.edu/Hippocrates/hippooath.html>

Hippocrates also appears to have sought out and met Democritus, possibly while they were both in Athens. One night at dinner together, Democritus ordered a cup of milk. When he received it, he pronounced that the goat the milk had been taken from was black and had delivered one kid. He was right. When Democritus saw Hippocrates one day with a maidservant, he greeted her with the words "Good morning, Maiden." The next morning, he saw the pair again and greeted the woman with "Good morning, woman." Somehow, Democritus was able to tell that Hippocrates had seduced her during the night.

Hippocrates wrote several books during his lifetime. In one of these works, entitled *On Airs, Waters and Places*, he cautions physicians to be aware of the area where they are diagnosing causes of illness and to notice which direction the prevailing winds are coming from, what the water supply is like in an area, etc. In addition, he has them take into account the proclivities of the local population and how much or little they ate or drank. After all, knowing these factors can help a physician properly determine the root cause of an illness:

"Whoever wishes to investigate medicine properly, should proceed thus: in the first place to consider the seasons of the year... Then the winds, the hot and the cold, especially such as are common to all countries and then such as are peculiar to each locality. We must also consider the qualities of the waters, for as they differ from one another in taste and weight, so also do they differ much in their qualities. In the same manner, when one comes into a city to which he is a stranger, he ought to consider its situation... and the mode in which the inhabitants live and what are their pursuits, whether they are fond of drinking and eating to excess and given to indolence, or are fond of exercise and labor and not given to excess in eating and drinking."[1]

In his book "On Ancient Medicine," Hippocrates discusses how the food now consumed by people (and preparation methods, such as cooking) lead to greater health for people than the food eaten by "the ox, the horse and all other animals except man," but that these early foods made mankind a stronger race today:

"At first, I am of opinion that man used the same sort of food and that the present articles of diet had been discovered and invented only after a long lapse of time, for when they suffered much and severely from strong and brutish diet, swallowing things which were raw, unmixed and possessing great strength, they became exposed to strong pains and diseases and to early deaths.... [T]hose who had weaker constitutions, would all perish; whereas the stronger would hold out for a longer time... [and}get off with little trouble, but others with much pain and suffering."

[1] Hippocrates. "On Airs, Waters and Places," Part one, circa 400 BCE. Reprinted in the MIT Internet Classics Archive, http://classics.mit.edu/Hippocrates/airwatpl.1.1.html.

Hippocrates was also critical of medical practitioners, saying in "The Law," "Physicians are many in title but very few in reality." Furthermore, he tells of what he believes are the qualifications and the personality traits which are required for a good physician:

"Whoever is to acquire a competent knowledge of medicine, ought to be possessed of the following advantages: a natural disposition; instruction; a favorable position for the study; early tuition; love of labor; leisure. First of all, a natural talent is required; He must also bring to the task a love of labor and perseverance, so that the instruction taking root may bring forth proper and abundant fruits."

Hippocrates also wrote several other books, including "On Fractures," "On Injuries of the Head," "Of the Epidemics" and "On the Surgery" (a step-by-step guide for people learning general surgery). "On the Epidemics" concerned one epidemic on Thasus (or Thasos), an island in the northern Aegean, using various case studies in order to show the tools of diagnosis, both while the disease was occurring and post-mortem.

When Hippocrates was thirty-three years old, Sophocles released his newest play, "Oedipus Rex." This famous play revolves around a son who is destined by fate to murder his father. When told of this by a prophet, the parents decide to abandon their son on a hillside to die. Raised by a shepherd, Oedipus comes upon his father (who he does not recognize) along the road and the two quarrel. At the end of the battle, Oedipus has fulfilled his destiny. However, the story is not yet over as Oedipus goes on to solve the mystery of the Sphinx. In reward for this accomplishment, he is awarded the hand of Jocasta in marriage. This woman, unbeknownst to either, is the mother of Oedipus. The morality lesson in the play being that although Oedipus had been fated to kill his father, the compliance of Jocasta in the attempted murder of their son added yet another curse onto the head of Jocasta: incest.

Although Hippocrates, like Empedocles before him, believed that the body contained four humors, Hippocrates believed outside stimuli were responsible for imbalances between them. He also taught that diet and hygiene were of utmost importance in the prevention of disease and that medicine should only take over where those fell short. Hippocrates taught that a physician should carefully observe his patient and about the health benefits of being in nature. He also spoke of traits being passed down along family lines and diseases being passed down over several generations, foretelling genetics, which would not be truly understood until Mendel, in 1866.

One interesting fact is that the Greeks (especially after Hippocrates) practiced very clean surgery and infection was uncommon. Hippocrates himself used (and taught the use of) boiled water or wine to irrigate open wounds. He brought new terminology into medicine, which is still in use today, including the terms: *crisis, relapse, exacerbate, chronic and convalescence.*

Hippocrates died in 370 BCE, on Cos, where he had been born ninety years earlier. Medicine would have to wait over four hundred years until another doctor came along who would be nearly as great as Hippocrates. At the time of the death of Hippocrates, the second of a line of three teachers and students was entering his late fifties. After this line, one student of the third teacher would nearly take over the world. The first of these instructors was Socrates, who taught our second teacher, Plato. Plato would go on to teach Aristotle and after he left Plato, Aristotle would give instruction to a young prince named Alexander. Socrates never much cared for science, but his student, Plato, an ex-wrestler and poet, certainly had a love for sciences.

Plato 427 - 347 BCE

Plato, from a 19th century woodcut.

Public domain photo

One of the most famous of all the ancient Greek philosophers, Plato was born in Athens about 427 BCE, where he would die eighty years later.

Four years before he was born, the rivalry between the maritime-centered Athens and the land empire of Sparta had broken out into the Peloponnesian war. As leader of the burgeoning democracy of Athens, Pericles supported democracies throughout the region. However, Sparta supported oligarchies similar to their own system. The war that resulted lasted twenty-seven years, with Athens winning most of the sea battles and Sparta winning most land battles. That was, of course, until Persia financed a fleet of ships for Sparta, under the command of Lysander, whose presence produced an uprising in Athenian territories. Two years before the birth of Plato, plague tore its way through Athens, killing over one-third of the population of that city, including Pericles himself, in 429 BCE.

Plato was born to Ariston and Perictione and his family traced their lineage back to the family of Solon the lawmaker. His mother, Perictione, was five generations removed from Solon's brother, Diopidas. He had two brothers, named Adimantus and Glaucon and a sister by the name of Petone.

Plato's name in childhood was Aristocles and the name Plato was given to him in school, where he was known as Platon, meaning broad, due to his broad shoulders. Taking advantage of the breadth of his body, he had been a wrestler in school, learning under a professional wrestler named Ariston of Argos. Despite the great width of his body and his stoutly build, the voice of Plato was said to be high pitched and had a sweet melody, which kept his listeners enraptured. Even as a child, he never laughed greatly, only in small, quiet chuckles. This lead people to believe he was solemn, but this trait is likely just a consequence of a taciturn personality.

He was first interested in poetry, writing tragedies and lyric poems and he also painted, directing his creative abilities into art. In 409 BCE, when he was eighteen, the talented youth became a devoted follower of the philosopher Socrates and the two became teacher and student. Plato was so enamored with the writings of Socrates, that despite being eligible for an award in tragedies at the Theatre of Bacchus, he stopped writing his own poetry and devoted himself instead to learning the poetry of Socrates.

Socrates had been born in Athens in the year 470 BCE, the son of a sculptor and a midwife. He was a physically unimpressive man; he was stocky in appearance, but was known for his great self-control and he rejected wealth as a path to a happier life.

He was schooled in music, gymnastics and literature. Socrates learned under the guidance of Anaxagoras, until the time when Anaxagoras was imprisoned, at which point he went to Archelaus for further education. Archelaus had worked at trying to reconcile the divergent ideas from Ionia and Athens, a series of debates that played a dominant role in thinking at this time.

In his youth, Socrates became a soldier, being recognized for bravery on more than one occasion. Plato had also been a soldier and was recognized once for valor during combat. This was largely the extent of all the traveling that Socrates would partake in during his life, preferring instead to remain at home.

Socrates also thought he could hear a voice in his head (perhaps he did). This would lead to trouble for him later. He held himself away from politics, since the voice told him to keep teaching. Listening to the voice, as he would work the streets of Athens, arguing and pontificating to anyone who would listen.

He seemed to be a generally happy person and had many friends, who enjoyed his clever wit and humor, which was said to be nearly free of cynicism or satire. When in a heated discussion, Socrates would often feign ignorance, in order to encourage the other person to restate their position more clearly, to learn more about their position, or to expose their argument as fallacious. This became known as "Socratic irony," or *daemonion*. Plato later claimed that Socrates "hid" behind this method of discourse.

Socrates, the famous philosopher who taught Plato

Public domain photo

Many citizens mistrusted Socrates and the humorist Aristophanes (in "Clouds") portrayed him as the leader of the Sophists, a group that taught young people that truth and morality had no absolute standard and how to make weak, incorrect reasoning appear powerful and logical. In addition, they took money for their teaching, which was frowned upon in early fourth century BCE Athens.

Socrates was thought to be dangerous for being aloof from the political process and, in 399 BCE, he was taken from his home in the middle of the night and charged with several crimes. These included ignoring the Gods and introducing his own deities (in fact, it was his "inner voice" which brought about this charge), as well as corrupting the youth, leading them away from democracy and he was (incorrectly) charged with being a Sophist (a group that he, like Plato and Aristotle after him, argued against). In fact, one of the students of Socrates actually began to take money for teaching and sent some of it to Socrates, who returned the money, claiming that the voice in his head had told him not to accept the donation. The charge of being a Sophist may have been attributable to his portrayal in "Clouds."

According to Plato, at the trial of Socrates, the defense of the accused was a wide, bold vindication of his entire life, placing the value of his life to the city as his sole defense to the charges presented. At age seventy, he claimed that if the Democrats put him to death, it would harm the city more than it would himself. He was condemned to death, but only by a small majority. When, as Greek law provided, he was allowed to present a counter-proposal, he only offered to pay a small fine and the assembled crowd voted again. This time, they voted for execution by an even wider margin.

Friends of Socrates planned an escape from jail for him, but he chose to comply with his sentence. The last day of his life found him surrounded by friends and admirers. Come the evening, he calmly followed the traditional method of execution, consuming a drink prepared from hemlock,* likely suffering for hours before finally giving in to the effects of the poison, leaving behind no writings and having never opened a school.

Socrates certainly did have disciples, however. One of these followers was named Xenophon, who came from Athens and was forty years younger than Socrates. Two years before the execution of Socrates, Xenophon had joined a group of Greek mercenaries who were taking part in a Persian civil war between two brothers fighting for power. After being defeated at the battle of Cunaxa, he led his weary troops back to Greek territory through 2,400 kilometers (1,500 miles) of swamps, famine, supply failures and many other hardships. He wrote his best-known work about these events in his book "Anabasis."

He also wrote many other works, including "Hellenica," "Cyropaedia," "Memorabilia" and "Oikonomikos." The first two were works on history ("Hellenica" was a further continuation on a earlier work by Thucydides on the Peloponnesian War and "Cyropaedia" was a biography of Cyrus the Great). The third book was the recollections of Xenophon about Socrates, but the fourth book was the most far-reaching, although it was a treatise on how to conduct one's household.

The name *Oikonomikos* was derived from the word *oikos*, which means household and *nem* (changed in the word due to word usage), meaning to regulate or organize. It is from *Oikonomikos* that we get our modern word *economics*. This may seem strange to us that a book on how to run a household would have such a tremendous effect. However, this is easier to understand when you realize that a household in Ancient Greece or Rome was not the same as what we think of as a household today. The households in these societies consisted of a patriarchal head who held domain over everyone on the property (including his son's wives), as well as

---

* Hemlock is a dark green plant, with an unpleasant odor, related to parsley and grows up to ten feet tall. All parts of the plant are poisonous, containing coniine ($C_5H_{10}NC_3H_7$), a powerful alkaloid nerve agent. The ancient Greeks knew several poisons well, and their word *tox* (meaning arrow) is the root of our modern words toxic and toxicology.

all of his servants and slaves. It is this sort of grouping that the Romans meant when they used the word *familia,* from which we get our modern word *family*. Thus, this work on household management included the advice on the proper management of slaves, servants, farms and wealth and it is now known as one of the founding documents of the modern science of economics.

After the execution of Socrates by the Democrats in 399 BCE, the twenty-eight year old Plato quickly fled the city of Athens and traveled around the Mediterranean, visiting Megara (a declining maritime city in eastern Greece) and then on to Cyrene, an intellectual center in what is now northeastern Libya. This city was known for being a place where great research was being done in medicine and philosophy. Here he learned under the tutelage of Theodorus the mathematician. From Cyrene, Plato traveled to the Italian peninsula, where he met with Pythagoreans who had made that area their home and then he went to Eurytus, where he fell ill and was cured of his affliction by the application of salt water. This event would lead Plato to believe that salt water could cure all ills. He considered taking a trip to Egypt, where he was told that everyone was a physician. However, the Peloponnesian War that was occurring at this time made such a trip impossible.

Plato began to strike back against, at least, the Athenian government. He is quoted as saying that things would never run smoothly in society until "kings were philosophers or philosophers were kings"[*] and he is also quoted as saying that "Those who are too smart to engage in politics are punished by being governed by those who are dumber."[†]

Despite his opposition to tyrannical government, he seems to have profited from his relationship with Dionysius, the tyrant of Syracuse, who gave Plato over eighty talents, which was over a ton of gold, worth nearly $40 million dollars today. He had the wealth to pay scouts to travel a great distance, just to acquire three books by Pythagoras, likely spending far more in the cost of travel than the value of the books themselves.

Of course, things did not always go so smoothly between Dionysius and Plato. Plato believed that there were five types of government, democracies, monarchies, aristocracies, oligarchies and tyrannies. The last type was, by far, his least favorite and he had let Dionysius know that fact. When they first met during the first of Plato's three trips to Sicily, they had a heated discussion about politics, which involved Plato telling the tyrant that he did not believe that one man alone should rule over people unless that person was pre-eminent in virtue.

Dionysius responded that Plato sounded like "an old dotard," or a senile person. Plato's response to that was "And your language is that of a tyrant."[1] Dionysius was enflamed and wanted to put Plato to death for his insolence, but was stayed when his

---

[*] Taylor, Brian. "Plato." http://www.briantaylor.com/Plato.htm

[†] Week, The. "Wit & Wisdom." 5.195. 18 Feb., 2005.

[1] Diogenes Laertius. "Lives of Eminent Philosophers." *Plato.*

associates talked him out of the execution. Later, Dionysius sent a letter to Plato, ordering him to never speak ill of him again and Plato wrote back that he did not have enough leisure time to think about the tyrant.

After the death of Dionysius, his son (also named Dionysius) took over the leadership of Syracuse. Plato asked Dionysius the Younger for land and an army so that he might live according to his ideas of government and society. The younger Dionysius agreed to supplies these for Plato, but never carried through on his promise.

The leaders of Thebes and Arcadia began to found a new city, offering Plato the city's leadership. However, they would not agree to Plato's demand for equal rights for all people and he refused the title. He would stand for the rights of people, even when there was danger to himself in the act. His friend, Chabrias, was a general who was facing a charge that could lead to execution. When he and Chabrias were heading to the Acropolis for the trial, Plato was approached by a man named Crobylus, who had spent his life garnering favors from influential people. Crobylus asked him, "Are you come to plead for another, not knowing that the hemlock of Socrates is waiting also for you?" Plato replied, "When I fought for my country, I encountered dangers and now too I encounter them in the cause of justice and for the defense of a friend."[1]

Plato was also opposed to what he perceived as moral degradation, including gambling, excessive drinking (except at the festivals for the God of wine) and sleeping. Once, he approached a man who was playing dice, reprimanding him for the act. The man responded that he was only playing for a trifle. "But the habit is not a trifle." answered Plato.[1]

Plato founded his own school of philosophy in 347 BCE. This school, known as *The Academy*, which was free to the students and admitted women as pupils, was supported by philanthropists and was dedicated to the God Academus. The name Academy would soon be applied to all schools of higher learning. Above the entranceway to the Academy were written the words "Let no one ignorant of mathematics enter here."

Plato's Academy lived on long after the death of its founder. The school was still educating students until 529 CE, when it was ordered shut down by the Byzantine emperor Justinian from his capital in Constantinople. The longevity of his Academy made Plato the preeminent philosopher in the western world until the thirteenth century, when the views of Aristotle began to gain predominance.

Towards the end of his life, Plato became enamored of the idea of a statue or monument being left behind to honor his teachings. When an admirer asked him if such a monument would be constructed, as they had been for previous philosophers, Plato answered, "A man must first make a name and the monument will follow."[1]

Plato believed, incorrectly, in the perfection of the heavens. He put forth the proposition that the orbits of the planets must be perfect spheres. This idea would prevent an accurate prediction of the positions of the planets until Kepler, over 2000

---

[1] Diogenes Laertius. "Lives of Eminent Philosophers." *Plato*.

years later, rejected the notion of the perfection of the heavens. Perhaps Plato held an unrealistic expectation of perfection because of his disdain for the mechanical arts. He considered any mechanical studies to be merely works of war, unworthy of his attention. Many philosophers followed his guidance and mechanics now fell from the purview of philosophers to that of engineers. Physicists are very exacting in their works, but engineers make things happen. If he had more mechanical experience, Plato might have had more tolerance for making things work.*

He stated that the Earth was made up from each of the four elements, earth, air, fire and water, so that it would be visible (from fire) and firm (from earth) and that the air and water were present so that the other two elements were not in disproportion to the other two. He also believed that species could not adapt or evolve. Plato, like Aristotle, seems testament to the fact that even brilliant people can sometimes be wrong.

Despite believing that the Gods had created order out of chaos, he was still dogged by religious fundamentalists. It appears that the debate of evolution vs. creationism was raging even then in ancient Greece. Plato would not be the last of the ancient scientists to feel the pressure of religious intolerance – it would play a major part in the downfall of ancient science.

Plato died peacefully in his sleep after attending the wedding feast of one of his students and he was buried in the Academy. He left behind one son, Ademantus, as well as four slaves. Our next great scientist was someone who Plato felt was a rival of his: Eudoxus of Cnidus.

* There's an old joke in scientific circles: "What's the difference between a physicist and an engineer?" Answer: "About one part in ten."

Eudoxus of Cnidus 408 - 355 BCE

Eudoxus of Cnidus

Photo believed to be public domain

Eudoxus was born in Cnidus, the son of Aischines in 408 BCE. He was four years old when the Peloponnesian War between Athens and Sparta ended, with a ruined and defeated Athens*. Lysander, who had commanded the Persian-financed Spartan fleet, installed an oligarchy in Athens, under the control of a body titled *The Thirty Tyrants*. Athens would never regain its former glory and for thirty years, Sparta was the dominant military force in Greece.

Eudoxus traveled greatly in his time as an adult, visiting Tarentum (now part of Italy), as well as Sicily, Egypt and Athens. He would go on to do important work in astronomy and he was quite an accomplished mathematician, in addition to being a physician, geometer and legislator.

It was in Tarentum that he studied with Archytas, who was a follower of Pythagoras. It was likely from Archytas that he learned about number theory and early musical theory.

He also began work on a problem known as *duplicating the cube*. The problem is, essentially, trying to find a way of using a compass and ruler (the workhorses of Greek mathematics) to find the cube root of two. Eudoxus also contributed early work on the *method of exhaustion*, an early procedure to find the area under a curve in a graph. Today, that method has been supplanted by the use of integral calculus. Another mathematical problem that concerned Eudoxus was finding ways of comparing two lengths, using only whole numbers and fractions. The work he did on this subject was later written as definition four of the fifth book of Euclid's *Elements*.

Eudoxus then went on to study medicine with Philiston in Sicily. Throughout his career as a student, Eudoxus strived to learn different subjects from different teachers, in order to avail himself of the best knowledge of his day.

From Sicily, he traveled to Athens, the center of scientific Greek culture. He spent much of his time with the physician Theomedon and likely attended talks by Plato. While studying in Athens, he lived at the docks and would come into the city each day to hear the lectures. Plato later threw Eudoxus out of the Academy, for reasons unknown. This was likely the beginning of bad blood between the two philosophers.

After only two months in Athens, Eudoxus left for Egypt, where he studied with the priests of Heliopolis and worked at an observatory located between Heliopolis and

---

* Athens actually had two chances to win the war, with requests from Sparta for terms of peace, in about 426 BCE and in 406 BCE and turned down the offer each time.

Cercesura. Before leaving to study with the Egyptian astronomer/priests, he shaved his eyebrows, as was the custom among the Egyptian intelligentsia. The people of Egypt began to call him Endoxus, meaning *glorious*.

In his book "On Velocities," Eudoxus believed that the planets and stars were fixed to spheres, which circled around the Earth, as did Pythagoras. However, he attempted to explain the discrepancies between such an orbit and actual observations by proposing that the axes of the twenty-seven spheres rotated about themselves like a top losing speed. Even still, the planetary model of Eudoxus failed to meet up to a rigorous observation of the actual paths of celestial paths throughout the sky.

Eudoxus, however, appears to have believed that this model was just a mathematical convenience, rather than a model of the actual world, since he never spoke of the composition of the spheres themselves. Later, Aristotle would take this theory, but he would choose to describe it as a physical model, rather than as a mathematical one.

His next move was to Cyzicus (in the northeastern part of the Ionian world), where he established a successful school, which earned him many devoted followers. While here, he built an observatory and made studies of the star Canopus. The observations made here, as well as at an observatory near Heliopolis, formed the basis of two books by Hipparchus on astronomy.

As if he had been planning it all along, sometime around 368 BCE, Eudoxus returned to Athens, this time carrying with him a retinue of devotees, possibly for the purpose of annoying Plato. While here, Eudoxus first began to sit his students in a semi-circle around him to hear the lectures, a practice that is now seeing a revival in classrooms across the nation. Plato was likely unhappy with the size of Eudoxus' school and Eudoxus considered Plato's mathematics lacking (which holds more than a measure of truth!).

After his visit back to his old school, Eudoxus returned home once more. However, this time, he returned to Cnidus, the land of his birth. The people there elected him to a high post in the legislature. He spent the rest of his days writing on astronomy, meteorology, geometry and theology. His son, Aristagoras, was the teacher of Chrysippus, who was also preoccupied with the study of nature. Chrysippus went on to do research in eye surgery, one of the most delicate, difficult fields of surgery.

Eudoxus also wrote a series of seven books, entitled *Tour of the Earth*. This series of books was, in essence, a work of physical geography, telling of all the people and lands that were known to Eudoxus, including their political and social systems. In these books, Eudoxus tells at length and in great detail of the people of Egypt in the fourth century BCE.

The years surrounding the death of Eudoxus were witness to another end and another beginning. For one year before the death of Eudoxus in 355 BCE, a fire destroyed the Temple of Artemis at Ephesus. The temple, along with the later

Lighthouse of Alexandria, was one of the seven wonders of the ancient world. One year after the death of Eudoxus, in 354 BCE, another wonder of the Ancient world was constructed; the Mausoleum of Halicarnassus, which held the body of King Mausolus.

None of the works completed by Eudoxus remain, nor do the works by Hipparchus that were based on the work of Eudoxus. However, many later writers told of his work, including over one hundred writers who made mention of his work *Tour of the Earth*.

Meanwhile, philosophy and education were about to make further advances as a small, frail well-dressed man from Stagira studied at Plato's Academy: Aristotle.

A few pieces of columns that have been restacked are all that remain today of the ancient Temple of Artemis at Ephesus.

GNU Free documentation license photo

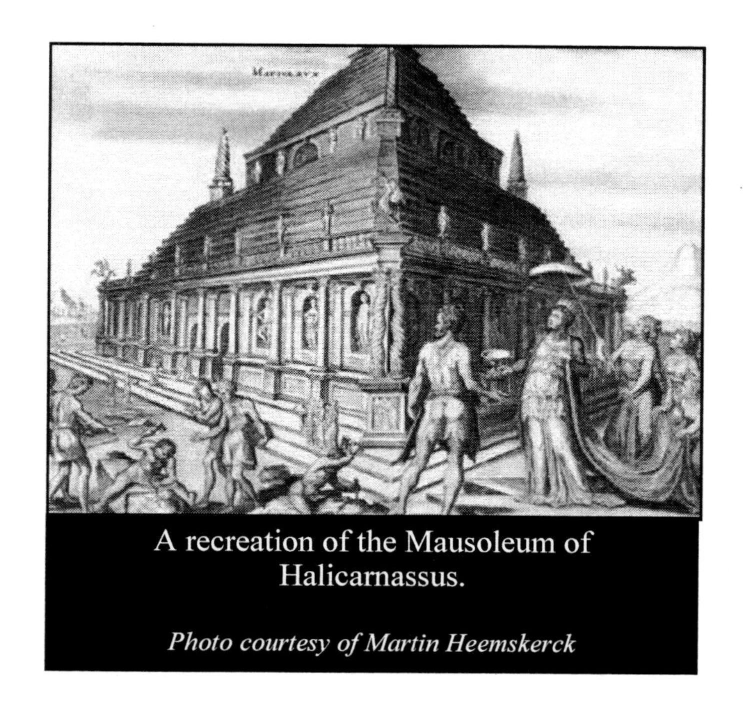

A recreation of the Mausoleum of Halicarnassus.

Photo courtesy of Martin Heemskerck

Aristotle 384 - 322 BCE

The most famous of all Plato's students was Aristotle. Aristotle was the ancient world's most prolific author and he would go on to found his own academy, the Lyceum. The Lyceum would remain strong throughout the Alexandrian period, although his work would long be forgotten.

Aristotle; from *The School of Athens* fresco by Raphael

Public domain photo

He had been born in 384 BCE, to Nichomachus and Phaestias, in the city of Stagira, in Macedonia. His father, Nichomachus, was a court physician and his mother was a resident of Stagira at the time Aristotle was born. That is about all that is known of Phaestias.

He seemed to be destined for a life of medicine, traveling the countryside with his father, treating the ill. In this age, knowledge of medicine was largely passed from father to son and had events not intervened, Aristotle would have likely become a physician himself, to carry on the knowledge and work of Nichomachus.

Aristotle was only ten years old when his father died and his mother died soon afterwards. He was raised by his uncle Proxenus and Proxenus began to teach the youngster about language, poetry and rhetoric. This artistic knowledge complimented the medical background that Aristotle had been raised in and played a large part in forming his personality as an adult.

It was in Aristotle's twelfth year that the Oracle of Delphi was destroyed. This was the best known and was considered the most powerful of the oracles of the ancient world. The oracle was located at the base of the southern slope of Mt. Parnassos, in Phocis, Greece.

When people came to ask questions of the oracle, a priestess would sit on a golden tripod (a three-legged stool) and become immersed in fumes seeping from the ground. A priest would speak in verse, translating her utterances, informing the questioner of the visions of the priestess. It is thought now that the perfume-like gases seeping from the floor was ethylene, which was used as a surgical anesthetic from the 1920s to the 1960s.

When he was seventeen, he left Stagira for Athens, where he joined Plato's Academy. Plato was not there at the time of Aristotle's arrival (being on a trip to Sicily at the time) and Eudoxus was heading the school in Plato's absence. Aristotle would stay at the Academy for twenty years, first as a student and later as a teacher.

At the time that Aristotle joined the Academy, Macedonia was in turmoil, suffering external wars, internal rebellions and several people held the throne over a short period of years. Finally, Macedonia found peace when Phillip II took the throne.

However, in so doing, the army of Phillip II (who may have been a childhood friend of Aristotle's) destroyed Aristotle's hometown of Stagira.

As an adult, Aristotle had a lisping voice, frail legs and small eyes. He dressed in the finest clothes, wore fine jewelry and always kept his hair well styled. He would have two children in his life, one daughter and a son named Nichomachus, after Aristotle's father. The mother of the younger Nichomachus was Aristotle's concubine, named Herpyllis.

Aristotle was a great believer in reason and he claimed that there was great regularity and order in nature. However, he also believed that observations were superficial, he failed to see the wisdom of using mathematics to uncover the laws of nature and he leapt to hasty generalizations.

He perfected syllogism and deductive reasoning. Syllogism is a system of logic which consists of a major premise, a minor premise and a conclusion. One of the most well-known examples is as follows:

1) All humans are mortal. [Major Premise]
2) I am a human. [Minor Premise]
3) I am mortal. [Conclusion]

Such reason (as in this example) can extend from the general (all humans are mortal) to the specific (I am mortal). This is known as deductive reasoning. Nevertheless, for whatever reason, Aristotle never glimpsed the logic and reason of mathematical approach to studying nature.

Aristotle taught that matter consisted of only one basic material and that the differences we observe in materials occur because of differing proportions of the qualities of dry vs. wet and hot vs. cold. These would then produce the four "elements" fire and air, earth and water.

In astronomy, Aristotle believed that each of the planets was carried on crystal spheres, rotating every day around the earth and that the outer spheres affected the motion of the inner spheres. This model universe of Aristotle was spherical, finite and closed. The middle area, nearest the Earth, contained the four elements, earth, air, fire and water. Each of these elements, he believed, rose or sank to their natural places: for instance, fire wanted to go upwards and earth wished to travel downwards. Once there, he taught that the elements would be at rest and stop moving. Since all these earthly ingredients traveled in straight lines (or so he believed) and the planets move in circular patterns, he proposed an additional element, which he dubbed "aither," written today as "ether."

The planets seem to change speed and even regularly travel backwards for periods of time as viewed from the Earth. This apparent backwards motion is called retrograde motion and occurs when one planet overtakes another and passes it in their tracks around the Sun. In an effort to explain these strange motions and always one to see regularity and order in nature, Aristotle designed a cumbersome system of fifty-six spheres to carry five planets, the Moon and Sun and the sphere of "fixed" stars.

Aristotle believed, again incorrectly, that that heavy objects fall faster than lighter objects (provided they had the same shape) and that the Earth was eternal, any signs of decay being balanced by rejuvenation.

Already, in the time of Aristotle, the very foundations that supported science had begun to crumble. There was little money to be made in science, making it difficult for these early scientists to make a living. They (and their rulers) saw very little use for science in the everyday world. Perhaps if they had, an early technological society may have formed, forever changing what we know as history. The scientists did not have what we know as the scientific method* and many had little use for experiment.

Upon the death of Plato in 347 BCE, Aristotle moved to Assos, in Asia Minor (near the island of Lesbos), where he married Pythias, the niece (and adopted daughter) of Hermias, a friend of Aristotle's, who ruled the city. The daughter of Aristotle was born from this marriage. His relationship with Herpyllis only came after the death of Pythias.

In 345 BCE, the Persians invaded Assos and Hermias was killed. Aristotle ran to safety in Pella, the capital of Macedonia. There, he began to teach Alexander, the son of the Macedonian leader Phillip II. Phillip also likely wanted to place Aristotle as the head of the Academy, since the leader of the school at that time was encouraging his students to oppose the rise of Macedonian power. When the child grew older, Aristotle made a request of Alexander to have Stagira rebuilt and his request was granted.

Aristotle returned to Athens in 335 BCE, when his former student, Alexander of Macedon, became king. This upstart eighteen-year-old would have great trials and accomplishments in his future, as we shall soon see.

It was during this stay in Athens that Aristotle founded the *Lyceum*, establishing his own school. The formation of the school may have been encouraged by Alexander, who saw it as a rival to the Academy. This became known as the Peripatetic school, meaning "walking school" due to the habit of Aristotle of walking around the grounds as he taught the classes.† At the Lyceum, Aristotle encouraged his students to ask him any questions that occurred to them, in contrast to the teaching style of Pythagoras.

---

* The scientific method is how science is (supposed) to be done today. A researcher observes a phenomenon, formulates a theory and then performs an experiment to independently test the data. When done correctly, the experiment will not bias the results towards one conclusion or another and the researcher will independently measure the data, coming to whichever conclusion fits the data, even if that is not what they had originally believed. This method is followed far too rarely, even today.

† The name Peripatetic may also have been due to an event when the young Alexander was sick and Aristotle walked around with him outside while they talked about nature, philosophy and science.

His students were also allowed to go at their own pace, pressing the people ahead of them for greater knowledge and not being bound by those who were slower than they. Unlike many of the ancient philosophers, he believed that disseminating knowledge to all people was the highest calling for an intellectual.

He was always devoted to education, telling people that the difference between an educated and an uneducated person was as great as the difference between the living and the dead. Aristotle believed that education was "an ornament in prosperity and a refuge in adversity."[1] and that education was the best thing a person could take to the grave with them.

One time, a fellow in the class was heckling Aristotle and he received no response from the philosopher. The heckler yelled, "Have I not been jeering you properly?" Aristotle's response was "Not that I know of, for I have not been listening to you."[1]

He wrote works on several subjects, including natural sciences, philosophy, rhetoric, poetry and politics. His book "Physics" told of his theories on astronomy, botany and meteorology. However, unlike Plato, Aristotle's main area of interest was biology, possibly inspired by his physician father.

In the field of biology, he believed that low life forms (such as flies and worms) could spontaneously generate from inanimate matter such as rotting food or dung. However, higher beings, he taught, all reproduced true to their original form. Therefore, he never believed in evolution, simply in the repeating life cycles of higher beings.

He taught that nature had a purpose and that all individual beings (which he extended to inanimate objects) had an innate desire to grow towards perfection. This led him to the mistaken belief that the reason objects fell to the ground was a desire on the part of that object to be closer to the Earth.

Aristotle was the first person to begin to categorize and separate the fields of science into several categories, that we would today call physics, astronomy, chemistry and so forth. In doing so, he lead the world from a composite of philosophy (which included science) into an era of specialty knowledge.

However, he also espoused the idea that scientists were folly to try to choose either empirical or deductive reasoning solely in their theories and should try to meld the two ways of thinking. This is a concept that today allows scientists such as Albert Einstein and Stephen Hawking to develop bold new theories and then prove them.

Einstein was famous for his thought experiments (like what would happen if you were in a free-falling elevator), which he would later show mathematically. Hawking is perhaps most famous for the discovery of "Hawking radiation," which is the only way we know of that anything can escape the grip of a black hole. He first thought of this while spending a great deal of time getting into bed, but was able to later show the proof of it through the use of mathematics. However, Aristotle never

[1] Diogenes Laertius. "Lives of Eminent Philosophers". *Aristotle.*

seemed to grasp the relation between empirical data and mathematics, although he was quite skilled in the science of numbers.

A student of Aristotle's Callippus of Cyzicus (370-310 BCE), went on to become the foremost astronomer of his day and had worked with Aristotle in Athens around the year 330 BCE. He was the first scientist to work on the geocentric (Earth centered) universe of Eudoxus. Callippus became aware of the model from another teacher, Polemarchus, who in turn, was a student of Eudoxus.

He carefully measured the length of the seasons, developed theories about comets and was the first to realize the unequal lengths of each of the seasons. To explain this discrepancy, he added six spheres to the planetary model of Eudoxus, which allowed the Sun to change velocity in its path around the Earth.

Stoic philosophers tended to distinguish themselves from other people by wearing the Abolla, a toga-like garment, made of wool, pictured here.

Public domain photo

He also produced a calendar to coordinate lunar and solar time scales. This Callippic cycle, as it is known, fit 940 lunar months into 76 tropical years, producing the far more accurate 365.25 days per year (it is that extra 1/4 day a year which causes leap years in our modern calendar). This system was so accurate, it was used for centuries by other astronomers.

This era also saw one of history's most colorful characters, Diogenes of Sinope. Full of vim and vigor, he was brimming with contempt for authority and convention.

Diogenes was born the son of Icesias, a local banker. Legend tells that either Icesias, or Icesias and Diogenes were found defacing coins in public. His father was imprisoned for the crime and Diogenes fled to Athens, where he appeared upon the doorstep of the Cynic Antisthenes. Antisthenes berated Diogenes, ordering him to leave and may even have swung at him with a stick. Diogenes informed Antisthenes that he had no stick hard enough to drive him from his presence, as long as Antisthenes had anything worthwhile to say. Impressed with the response, Antisthenes invited him into his house and the two became student and teacher.

Antisthenes taught that social conventions should be largely disregarded and pleasure shunned and Diogenes adopted the principles of his new teacher with great abandon.

Diogenes, by John William Waterhouse, painted in 1882.

Public domain photo

There is also a story told that once, while under the tutelage of Antisthenes, Diogenes saw a mouse scurrying around and, for a while, he watched the mouse as it played and went about its business. Diogenes came to the conclusion that the mouse had an enviable lifestyle; no desire for material goods, never worrying about where to sleep, or being afraid of the dark. These are principles upon which he would guide his life.

Being concerned solely with practical matters, he wore a rough cloak, slept on bare ground and carried just a wallet and staff (and he only carried the staff later in his life). Using public places as his main source of habitation, he depended on handouts and charity from friends and strangers to pay for his food, which often consisted of lentils. He considered this "repayment" of their debt to him for his teachings.

Diogenes always had contempt for many of the other philosophers around him, regarding Plato's lectures as "a waste of time." He also considered Plato to be someone who would "talk without end." Plato seemed to have little respect for Diogenes, as well.

Once, Plato came into an area where he saw Diogenes eating a plate of figs. Diogenes offered Plato some figs and Plato ate the entire plate. Replied Diogenes, "I said that you might share them, not that you might eat them all up."[1]

Another time, Diogenes asked Plato for some wine and figs. Plato sent him an entire large jar of figs. Diogenes asked Plato "If someone asks you how many two and two are, will you answer twenty? So, it seems you neither give as you are asked, nor answer as you are questioned."[1]

In a large audience one day, Diogenes was listening to a speech by Plato, in which he defined man as "an animal… biped and featherless." The crowd erupted in cheers, except for Diogenes, who headed to the kitchen, where he plucked a chicken. He brought it out, declaring, "Here is Plato's man." Plato then added, "having broad nails" to his definition of what makes a man.

One time, it is told, Diogenes was seen in the city square, begging from a statue.

[1] Diogenes Laertius. "Lives of Eminent Philosophers." *Diogenes.*

When asked what he could possibly be doing, he replied that he was practicing being rejected. Another day, Diogenes was outside preaching philosophy and no one was paying attention to him. So, he began to whistle and a crowd gathered. Diogenes then reproached the crowd for coming to hear nonsense, but not serious discourse.

His barebones lifestyle led the people around him to give him the nickname *Kyon*, meaning "dog."* Antisthenes would often teach in a gymnasium named *Cynosarges*. It is likely from either *Kyon* or *Cynosarges* that we derive our modern word *cynic*. The famous school of philosophy known as cynicism was thought to have been founded by Diogenes or Antisthenes, probably both had a say in the early doctrines of this school of philosophy. Cynicism, as a philosophy, is more active in their disapproval of others who choose to gather wealth than stoics, but in many other ways the two philosophies are very similar. The teachings of Diogenes even inspired one of his wealthy admirers, Crates of Thebes, to give away all his earthly possessions and live a simple, austere life.

One day, while eating a bowl of lentils, Diogenes was happened upon by the philosopher Aristippus, who had become wealthy through his constant glorification of Athen's leader, the tyrant Dionysus. Aristippus informed Diogenes that if he would just learn to flatter Dionysus, he would not have to eat lentils. To this, Diogenes retorted that if Aristippus would just learn to eat lentils, he would not have to flatter Dionysus.

Another story has Diogenes approached by a woman, who complained to the philosopher that her son was rude and ill tempered. She asked him what to do about the situation. Diogenes slapped the mother.

Aristotle also knew Diogenes, but he also knew of the sharp wit of the impoverished philosopher. One day when they met, Diogenes handed Aristotle a fig. Aristotle knew that he could not refuse it without feeling the barbs of Diogenes' quick impudence. Therefore, Aristotle took the fig, telling Diogenes that he had lost both his joke and his fig as well. When Diogenes tried again later, handing Aristotle another fig, Aristotle took it, clasped it in his hands and imitated a small child, exclaiming, "Oh, great Diogenes" with exaggerated abandon. Then he handed the fig back to Diogenes, who was left speechless, which was a rare occurrence. Few people but Aristotle ever had the ability to leave Diogenes without a comment.

Traveling by ship one day to Aegina, Diogenes was captured by pirates and was sold as a slave. When the auctioneer asked Diogenes what skills he was proficient in, he replied "In ruling men." Pointing at a man in the audience dressed in a fine robe with a purple sash, Diogenes stated "Sell me to this man; he needs a master." The man in the assembled crowd, named Xeniades, bought Diogenes and made him the teacher

---

* The name Kyon was given to Diogenes by Plato. His response to this nickname was "Quite true. For I have come back again and again to those who have sold me." (Diogenes Laertius in "Lives of Eminent Philosophers." *Diogenes*.)

of his children. He told his friends “A good genius has entered my house.”[1]

One of the people who studied under Diogenes was named Crates and it was he who would go on to influence Zeno of Citium, who was partially responsible for the development of the school of Stoicism. However, Crates was not alone in his love for and devotion to, philosophy.

His wife, Hipparchia, was extremely devoted to contemplation, preferring philosophy to everyday life. In an age when many marriages were arranged, loveless affairs, she was also very devoted to her husband from before the time they were married, threatening to kill herself if her parents did not let her marry Crates. She was so enamored with her husband that she adopted his style of dress and they were rarely in public without the other by their side. When Theodorus berated her for obviously not attending to her loom, she replied “Do I appear to you to have come to a wrong decision, if I devote that time to philosophy, which I otherwise should have spent at the loom?"[2]

In 344 BCE, Aristotle went to the island of Lesbos, where he studied natural history, including a great deal of study in marine biology. For his trip over, for the first time in history, his ship may have been covered by maritime insurance – one of the oldest forms of insurance, which started about this time. Pirates were roaming the Mediterranean, taking ships and the contents of private vessels that were loaded with supplies or wealth. Since there was great wealth to be lost to these raids, as well as money to be made by insuring vessels, the great world of insurance was formed.

When his former student, Alexander, died in 323 BCE, there was a tremendous anti-Macedonian sentiment brewing in Athens and Aristotle quickly left for his family estate in Euboea, where he lived in a house that had belonged to his mother. He would live out the rest of his days here before dying one year later, after complaining of a stomachache.

During his life, he wrote dozens of books on subjects of science, philosophy, politics and more, totaling nearly half a million lines of text, but they soon disappeared. Very little remains of the philosophical of scientific works of Aristotle, except for one thing. As he lectured at the school, strolling around talking to the students, he was working from carefully prepared notes he had prepared for his lectures. These notes (totaling over 2,000 pages) survived long after Aristotle and it is from these notes, recompiled by later historians (including the Lyceum’s last director in 60 BCE), that we have most of our knowledge about Aristotle today.

These notes were largely by the west ignored until the thirteenth century CE, when Aristotle’s philosophical basis of the “perfection” of the heavens and of the repeating life cycles of higher being were used as evidence for Christian thought by St. Thomas Aquinas. This act was one of the major factors that would lead later to the Renaissance.

---

[1] Diogenes Laertius. “Lives of Eminent Philosophers.” *Diogenes.*

[2] Diogenes Laertius. “Lives of Eminent Philosophers.” *Hipparchia.*

Alexander the Great 356 - 323 BCE

Alexander the Great, from a mosaic found in the ruins of Pompei.

Public domain photo

Alexander the Great led a short, brutal life full of accomplishment and glory. The greatest general the world had ever seen (or perhaps has ever seen since), was himself the son of the Macedonian general, Phillip II.

Five years after Aristotle left for Lesbos, Macedonia came under the rule of Phillip II, who had ruthlessly grabbed the throne by force and suppressed his opposition in the year 359 BCE. Phillip had been a hostage in Thebes for three years, having being freed only five years earlier. While in prison, he learned much about Greece and its people.

After assuming the throne, Phillip quickly took to a program of expansion, using the tools of diplomacy and conquest to gain more land.

Phillip had multiple wives and favored another wife over one wife, Olympias, who was a princess of Epirus, an area of ancient Greece just south of Macedonia. This favoritism by Phillip for his other wives likely created resentment towards Phillip on the part of Olympias and her son, who had been born in 356 BCE.

As if to mark the end of an era, or the coming of another, the night Alexander was born, the first Temple of Artemis (dedicated to the Goddess Diana) burned to the ground. Legend has it that a man named Herostratus set the fire, in order to make a name for himself with one infamous act. After the temple burned, the city-state of Ephesus forbade its people from speaking the name of Herostratus, on pain of death. But a more likely explanation for the fire (given the heat necessary to ignite marble) is that the flames were started by a bolt of lightning.

The influences in his life included both artistic, as well as militaristic components. Likely, it was his mother Olympias who instilled in Alexander a love for art and an interest in mysticism, while his father was a general who conquered all of ancient Greece and encouraged the young Alexander to ride and live with him and his soldiers from the time he was a child.

A more copasetic man on the home front than on the battlefront, Phillip II gave his child the greatest teacher that money could buy. He had his young son, Alexander, taught by none other than Aristotle himself.

In 338 BCE, when Phillip had conquered Greece at Chaeronea, the final decisive maneuver was a cavalry charge, led by the eighteen-year-old Alexander. This was the first of many victories in battle for Alexander. He would soon become known for his out-of-the-box thinking and his cunning against nearly impossible odds.

In 336 BCE, while preparing for a conquest of Persia, Phillip II was murdered, which left Philip's empire in the hands of the 20-year-old Alexander. In the next thirteen years, Alexander would go on to rule most of the known world and would become one of the great military leaders of all time: Alexander the Great.

Olympias was accused of the murder of Phillip, although she may not have committed the act. Found innocent, she would soon get herself into deeper trouble. She was a violent woman and two decades after the death of Phillip, she opposed Cassander, who was regent in Macedonia in 316 BCE. He ordered her execution that year. She had outlived her son by seven years, but the legacy of her child with Phillip would ring down through the ages, to our present day.

Upon the ascending to the throne by Alexander, the region of Thessaly* had revolted against the young king and rebellion grew in his own kingdom. Thessaly, a large pastoral area about 13,000 km (5,000 miles) in area, wanted independence from the Macedonian rule that they had lived under since they were conquered by Phillip II. Moving quickly, Alexander ordered the execution of political dissidents at home and he led his army to crush the rebellion in Thessaly, which they accomplished by the end of Alexander's first summer in power.

The next year saw three different rebellions that were put down by Alexander. In a series of lightning-fast victories, he crushed revolts against his rule by the Thracians, Illyrians and the city of Thebes.† After his victory at Thebes, he destroyed the city, sparing only temples and the house of the poet Pindar. He sold the remaining 8,000 or so inhabitants into slavery. After these events, the rest of the lands under Macedonian rule submitted fully to the rule of Alexander without protest or rebellion.

The young king heard tales of the great philosopher Diogenes and went to find the upstart Cynic. He found Diogenes, outside sunbathing. Alexander walked up to the reclining philosopher, standing between him and the Sun. He proudly announced "I am Alexander." Diogenes took one look at the young Macedonian and just as firmly replied "And I am Diogenes." Alexander asked the philosopher if there was any favor he might do for him, given his royal power. Diogenes replied to the king "Yes. Get out of my Sun." Alexander asked him "Are you not afraid of me?" Diogenes inquired of the young king "Why, what are you? A good thing or bad?" Alexander answered "A good thing." Diogenes then replied, "Who then, is afraid of the good?" Smiling, Alexander stated, "If I were not Alexander, I would wish to be Diogenes."[1]

Alexander began to assemble Greek troops for an assault on Persia, which began in the spring of 334 BCE. Crossing the Hellespont with 35,000 men (who were

---

* Thessaly was the land where Jason and the Argonauts were said to have lead from in their quest for the Golden Fleece, as well as being home to the "centaurs", mythological creatures which were said to be half man, half horse.

† The city of Thebes had previously been well defended by a group of 150 male homosexual couples, known as the Band of Thebes. Their valor and bravery in battle was astounding.

[1] Diogenes Laertius. "Lives of Eminent Philosophers." *Diogenes*. The order of the lines spoken by the two men may have been different than what is portrayed here, but it is logical and makes for a good read.

Greek and Macedonian soldiers led by Macedonian generals), he marched southward. As his army approached the site of the ancient city of Troy, at the river Granicus, they encountered 40,000 Persian regulars and Greek mercenaries. Legend tells that Alexander lost only 110 men defeating the numerically superior Persian forces.

In Asia Minor, he entered the land of Phrygia; a barren plateau where only grapes found a natural home and where fine marble was quarried. Centuries before, an oracle had decreed that the next person walking through the gates of Phrygia should become their ruler. A peasant stepped through the gates and became their king. His name was Gordius and he declared that the cart he drove in to Phrygia would become a shrine to Zeus. He tied the pole of the wagon to the yoke with a knot, having no visible ends, that was so intricate that no person could manage to untie it. It was said that whoever would manage to undo the Gordian knot would rule all of Asia.

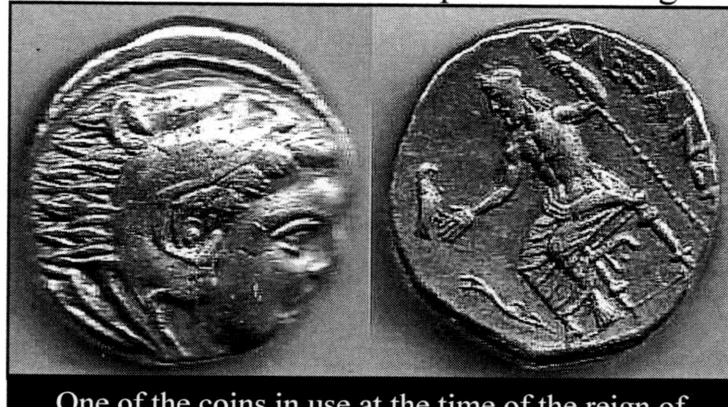

One of the coins in use at the time of the reign of Alexander the Great. On the front is a likeness of Alexander and on the back is Zeus sitting.

Courtesy Ancient Coins Canada

With Alexander being in Phrygia, he must have felt compelled to try his hand at this challenge that had baffled all comers for centuries; his penchant for mysticism and prophecy must have also compelled him to this task. Like all the people who had tried the challenge before him, he too was unable to find even the beginning of the puzzle. Undefeated, he rose and withdrew his sword. With a flurry of attacks upon the rope, he sliced the Gordian knot and the yoke and pole fell apart. Alexander was emboldened by this blessing and in practical terms, it also signified the out-of-the-box thinking that had served him so well against the Persian forces at Granicus and would go on to make even impossible battles winnable for Alexander as he nearly fulfilled the prophesy of Gordius.

He quickly won another victory against the Persians; it was at the battle of Issus in 333 BCE, in northeastern Syria, where he defeated a vastly superior force of Persians led by their king, Darius III. Defeated in battle, Darius fled to the north, leaving his wife, mother and children behind to be captured by Alexander. When Alexander found the royal family, he treated them well and ordered that they be given all the supplies they needed in order to keep living in the manner to which they were accustomed.

Next to the south for Alexander was a city which was said to immune from attack: the walled city of Tyre. This was a city, off the shore, built upon a small island.

But what made the defenses of the city so unique was the wall, which was built to the very edge of the water, so that an attacker could literally not even lay a step on the island, without getting through the wall first and this in itself was not an easy task. Alexander laid a siege upon Tyre, but the inhabitants would not surrender. Alexander's forces built a series of walkway bridges across the water, in order to enter the city. Repeatedly, the forces destroyed these bridges and with each bridge destroyed, they put off the invasion of Alexander for a little while longer. Finally, the bridges held and Tyre broke, with Alexander personally leading the charge over the wall. After seven months of effort, the city that could not be conquered fell to Alexander in 332 BCE.

This was going to be a big year for the young Alexander. In addition that year, he took Gaza and then stood with his army at the border into Egypt.

Egypt had been racked for decades by civil wars and wars with Persia, which had controlled Egypt for over a century before the arrival of Alexander. Under Xerxes I, the Persians had taken control of several Egyptian temples and revolts broke out across the Persian province of Egypt. Thirty years before Alexander arrived in Egypt, the Egyptian Pharaoh Nektanebo I was already discussing an alliance with Greece that would give Egypt a powerful friend in the fight against the Persians. Egypt went on the offensive against the Persians, led by the new pharaoh, Teos, the son of Nektanebo. Teos left to lead the army (which included Greek mercenary soldiers, especially from Sparta) while the Pharaoh's brother, Tjahapimu, ruled Egypt. Revolt broke out over the high taxes levied to pay for the war and during the civil wars which followed, Persia continued to invade. They finally gained control of all of Egypt for ten years, beginning in 342 BCE with the final defeat of Nektanebo II. The country was tired of war and unwilling to live under Persian rule. When Alexander arrived in the land of the pharaohs, he was welcomed with open arms as a deliverer of the Egyptian people.

Also in 332 BCE, Alexander founded the great port city of Alexandria, built up from a tiny fishing village called Rhakotis on the northern border of Egypt. It was here that the greatest library in the ancient world would be built and that institution of learning would make Alexandria the greatest center of learning in the world and the most cosmopolitan city in the Mediterranean.

Alexander took over Cyrene, the capital of the north African country of Cyrenaica, which brought his expanding empire to the borders of Carthage. Then the great conqueror spent six months wandering the Sahara desert. He and his entourage found themselves with little water and they needed it badly. As their journey continued, the need for water become ever more dangerous. Just as they reached the last of their supply of the vital liquid, a rare massive rainstorm developed in their area and they found themselves with more water than they could possibly want or need.

The following spring Alexander, always the mystic, went to gather the blessings of another soothsayer, this one at the temple of Amon-Ra in the Siwa Oasis. This

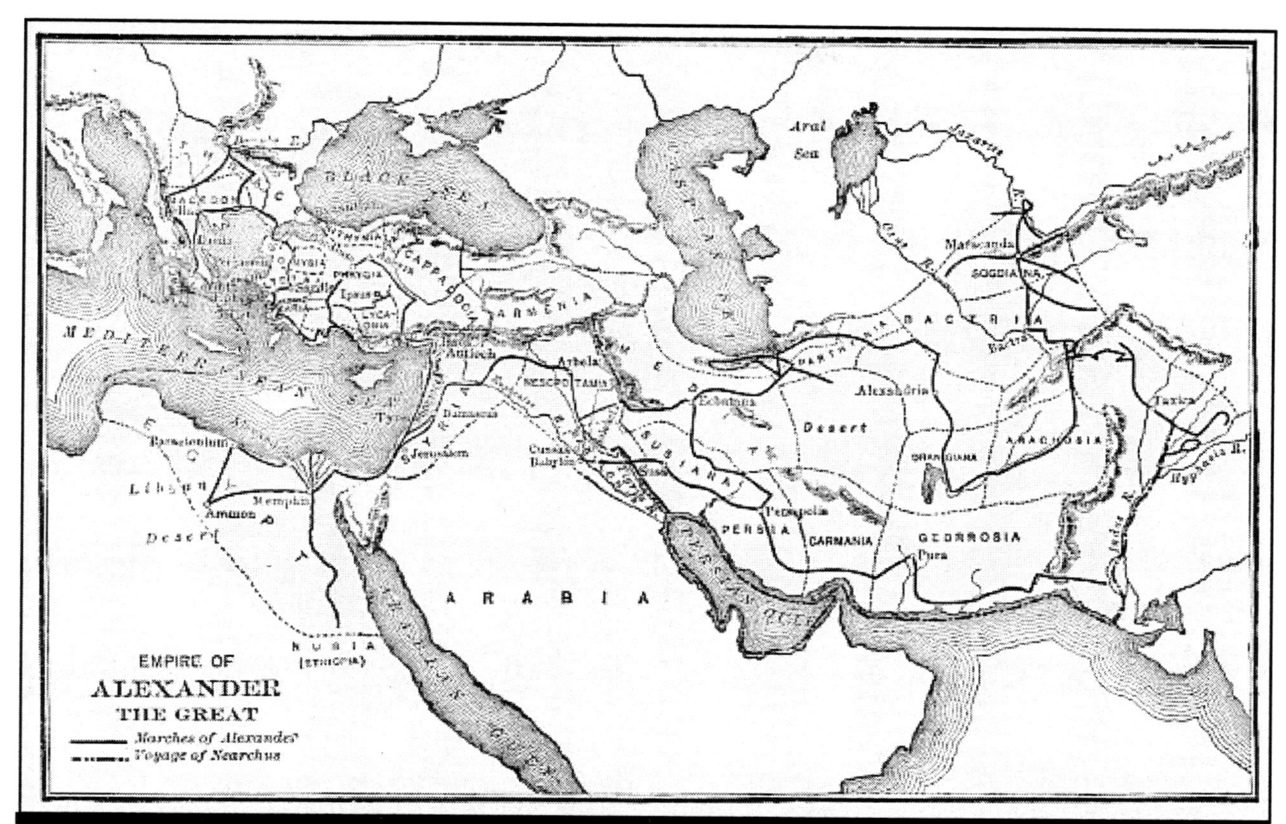

A map of the conquests of Alexander at the time of his death in 323 BCE, from the 1913 book *A History of the Ancient World* by George Willis Botsford, Ph. D.

Public domain photo

Egyptian God of the Sun was the parallel of the Greek god Zeus. Naturally, Alexander would seek the assistance of this most powerful of Gods and besides, the ancient pharaohs of Egypt were said to be the sons of Amon-Ra and getting this title as well would not be a bad political move for the young conqueror.

When he arrived at the temple, the temple priest, attempting to speak Greek (and not being very skilled in the language) meant to greet Alexander with the words "Oh, my son." Instead, the phrase he first spoke to Alexander was "Oh, Son of Zeus." It appeared to Alexander that he had been divined as the son of Amon-Ra/Zeus without much effort at all and he took this misspoken greeting as a sign of his divine lineage.

Alexander was allowed to enter the center of the temple alone; a privilege almost unheard of there. While inside, he stated that he asked three questions of Amon-Ra. The first question was "Would he rule the world?" to which he would tell the answer was "Yes." The second question was "Were all of the assassins of his father Phillip captured?" Again, the answer Alexander reported receiving was "Yes." The third question was "Am I the son of Zeus?" The answer, supposedly, was once again, "Yes."

Emboldened by the blessing of Amon-Ra and now convinced he was a child of

Zeus, he set back across the desert with his men, stopping first at Tyre to reorganize before then leading his 40,000 infantry troops and 7,000 cavalry on the long march to Babylon. It was after they had crossed the Tigris and Euphrates Rivers (in modern-day Iraq), in 331 BCE that Alexander once again engaged Darius III, at the battle of Gargamela on the first of October that year. Alexander's troops were, once again, far outnumbered and outgunned. Ancient sources state that Darius had a million men under his command at this battle, although this number is likely exaggerated.

Alexander's opponent also had the ultimate weapon of the day: the scythe chariot. These chariots, which featured one-meter long sword-like blades attached to the hubs of each of the chariot wheels, were thought to make an army nearly invincible. Had the opponent in battle against Darius been anyone else, perhaps this would have been the case, but not against Alexander. For the young general realized that horses would not be willing to strike into a line of infantry with large sharp pikes unless they had no other choice. Alexander gave the horses of Darius a choice. Interspersed among his line of pike men with their fourteen-foot spears, Alexander had groups of soldiers create "U" shaped openings with no pikes showing. Naturally, the horses leading the scythe chariots into battle chose these openings to enter. This was a deadly mistake. As they did so, the soldiers at the "bottom" of the "U" lowered their pikes, stopping the horses dead in their tracks. The drivers inside the chariots and the horses were surrounded on three sides by men wielding swords and pikes. This maneuver, known as a "mousetrap" (for obvious reasons), is still studied by military strategists in the modern day. The victory against the scythe chariots was so one-sided, that chariots were never again used as a weapon of war (to a major extent) anywhere in the world.

At the end of the battle, Alexander began to chase Darius off the field of battle, either in an attempt to kill the king and decapitate (quite literally) the head of the Persian armies, or to capture Darius alive, keeping him as a puppet governor in the Persian capital. After all, Persia at the time was filled with people who claimed to be the "true leader" of various areas and having Darius alive would keep a power vacuum from emerging, which would quickly fractionalize Persia, degrading the land into a system where powerful warlords would control small areas.

News came to Alexander while he was pursuing Darius that one of his flanks was losing badly to the Persians and that only the presence of Alexander himself could save them. He had a clear-cut choice; either he could catch Darius, thereby sacrificing a good portion of his army, or he could save his men, but the price of that would be the escape of Darius. Alexander chose to save his men, letting Darius escape. Within a year, two generals of Darius, participating with other conspirators within his government would arrest the leader and take command of Persia. At first cordial to Darius, the mood turned ugly when Darius refused to endorse their rebellion. One of the generals who led the revolt stabbed Darius to death for his refusal to cooperate.

Ironically, one of the last people to see Darius alive was none other than

Alexander himself, who came across the mortally wounded king. Alexander comforted Darius and gave him water. After Darius passed away, Alexander grieved for his fallen nemesis, gracefully covering the corpse with a blanket. Later, he had the body of Darius sent to his mother for burial. Alexander then tried to hunt down the two generals who assassinated their leader and caught one of them. Alexander ordered the assassin's body tied hand and foot to two sturdy young trees that had been bent for the purpose. The bent trees then had their ropes cut and the body of the assassin was torn in half and strewn apart along the ground to rot in the desert. This was the real dichotomy of Alexander. Sometimes gracious, loving and even jovial, he could turn in a second to cruelty, barbarism and sadism.

While on the campaign in Persia, Alexander had created dissention among the Persian ranks by enlisting many Persian soldiers to come to fight for him. No reason anyone should fight for Darius against the greatest general the world had ever seen, when that general was offering you a job, security and the fortune to most likely be on the winning side after the battle. In addition, Alexander had ordered, bribed and cajoled his officers to each take Persian brides, in an effort to further enhance the stability of his dominance over Persia. After all, most people are less likely to fight against an army in which a relative is a member.

It was after the battle at Gargamela that Alexander of Macedon took the traditional title of the Persian kings, becoming thereafter known as "Alexander the Great." Soon thereafter, Babylon surrendered to Alexander without a fight and the fabulously wealthy city of Susa easily fell to the Macedonian general quickly afterwards.

Eager to try to keep control of his vast empire, Alexander became convinced that the only way to prevent revolutions and dissent within his extensive lands was to win the hearts and minds of the people he had conquered. To this end, he had his men begin the reconstruction of temples in Babylon that had been destroyed earlier by Xerxes. This project would never see completion.

Persepolis, the capital of Persia, was next on Alexander's itinerary. His troops took the city and raided and plundered every treasure they could find within her borders. Always a violent drunk, one night, Alexander burned the city to the ground in a drunken rampage and with the destruction of Persepolis, the ancient Persian Empire was no more.

Alexander spent the next three years fighting and winning wars in all of Central Asia, including the areas that are now Afghanistan and Turkistan. Turkistan was an important crossroads in east/west trade and control of that area was critical for anyone who wished to control the trade between the borders of Alexander's now vast empire. He may have had dreams of uniting both east and west under a world government, with himself in control, of course. This area of Central Asia was critical.

It was while he was gaining control of this area that Alexander killed his friend Clitus in a drunken rage during an argument in 328 BCE. It was something he went into deep remorse and regret over. That incident was an act for which he may never have forgiven himself.

The following year Alexander met Roxane, the sixteen-year-old daughter of a Bactrian nobleman, who was said to be the most beautiful woman in all of Asia. Reportedly, it was a case of love at first sight and the two quickly married. One member of the wedding party was none other than Hephaestion, Alexander's male lover and close friend.

India had previously been a part of the Persian Empire and Alexander saw dominion over India to be necessary to ensure that Persia would never again rise as a world power, threatening his own control over the world. He had also heard tell of an "outer ocean" (the Pacific) that lay beyond the eastern coast of India. Alexander was curious enough where that alone may have driven him all the way to the east, even without the great wealth and power which would be gained by taking control of the Indian subcontinent.

In 326 BCE, he left Hephaestion in charge of Persia and Alexander and his troops crossed the Indus River, into the region of Punjab in northwestern India. Alexander's armies then continued sailing down the Indus River, conquering several cities on his way. However, the Macedonian troops had experienced enough fighting and they began to rebel against the young leader. In only a few short years, Alexander conquered much of the known world and was well on his way to conquering India when he headed back under threat of revolt and on the advice of his generals.

Assembling a larger fleet of ships, he continued to sail down to the mouth of the Indus and around western India to the Persian Gulf. While here, the troops carefully mapped the area for further future excursions in the region. Also in 323 BCE, Alexander married his second wife, Stateira, the daughter of Darius.

Filled with his triumph and glory, Alexander demanded that his troops kiss his hand. This was a Persian tradition, reserved only for people considered to be of divine birth and Callisthenes, a historian and nephew of Aristotle, refused to follow. He questioned Alexander, "Will you make the Greeks worship you as a God?"[1]

It turns out this is *exactly* what Alexander had in mind. That year, he gave the order that all the cities in Greece would worship him as a God. This order was probably due to a combination of believing that he was of divine birth (due to the events at the Temple of Amon-Ra) and for political reasons, but it caused great dissent among his followers.

The troops began to march overland towards the country of Media, which was located in what is now northwest Iran. The march was terrible and plagued by famine and drought. Spring 323 BCE saw Alexander back in Babylon, where he developed a plan to make Babylon and Alexandria twin capitols of his world empire. The arts, literature and science (with the exception of astronomy) that had become lifeless and derivative under the former Persian rule of Mesopotamia once again flourished in the

---

[1] History Channel, The. "The True Story of Alexander the Great."

years following Alexander's conquests.

In June of that year, Alexander contracted a fever (likely malaria) and died on the thirteenth of that month in Babylon, after suffering in bed for twelve days. There is also the possibility that he was poisoned by either conspirators, medical incompetence or both. Alexander lived to be just one month shy of thirty-three years old, but incredibly, by that age, he had managed to conquer nearly the entire known world. Even today, he is still regarded by many as the greatest general who has ever led an army into battle. Amazingly, through everything that it took to take over most of the known world, Alexander never lost even a single major battle.

On the very same day that Alexander died in Babylon, Diogenes died in Corinth at age ninety, possibly from eating a tainted raw octopus, or being bitten by dogs as he was feeding them octopus meat.

When his friends came across his body, they argued so much over who would have to bury him, that they came to blows. The city fathers finally had Diogenes entombed near the gates to Corinth and the people of the city created bronze statues to his memory, inscribed with their thoughts:

> *"Time makes even bronze grow old;*
> *But thy glory, Diogenes, all eternity will never destroy.*
> *Since thou alone didst point out to mortals the lesson...*
> *Of the easiest path of life."*[1]

At the time of the death of Alexander, Roxane was pregnant with their first child, whom Alexander never saw. This child would have been known as Alexander IV, however, he would never reach adulthood. The kingship of Macedonia was given to Alexander's mentally retarded brother Phillip Arrhideus, but his power was weak and unsure. The generals who had fought under Alexander throughout his campaigns wielded the true power in Macedonia. Roxane killed Stateira, fearing that she might be a rival to her and her son. Thirteen years later, both Roxane and Alexander's only son were killed after a long imprisonment by Cassander (married to Alexander's sister Thessalonica), who had become the king of Macedonia. This was the same man who had ordered the execution of Olympias in 316 BCE for the murder of Alexander's father, Phillip II.

After the death of Alexander, his body may have been interned for a while in the very same sarcophagus that was built to hold the body of the former pharaoh Nektanebo II. The same sarcophagus was also later used (with holes drilled in the bottom) as a ritual bath at the Attaria Mosque that was built in Alexandria over a thousand years after Alexander. Later, the sarcophagus/bath was taken by the British along with the Rosetta Stone* to the British Museum in 1800.

---

[1] Diogenes Laertius, in "Lives of Eminent Philosophers." *Diogenes.*

* The Rosetta Stone contained the same text written in Hieroglyphics, Coptic (the language of a

The year 323 BCE saw not only the death of Alexander, but also the retirement of Aristotle, who left the island of Lesbos and went to Chalcis (otherwise known as Khalks), where he would die one year later. Upon the death of Aristotle, the management of the Peripatetic School at the Lyceum was left to a student named Theophrastus, who would teach there for the next thirty-five years. This period saw a great peak of popularity for the Lyceum, as the attendance of the school passed 2,000 students. During that time, Theophrastus would produce works on geology, meteorology, physics, ethics, politics, poetry and more. He is well known for criticizing many of the theories of Aristotle (which in itself, is not such a bad thing!), but there were several models he developed which are of special interest.

One work of Theophrastus, entitled "On Stones," classifies the types of minerals known to him according to such factors as hardness, color, weight, solidity and so forth. These are some of the same qualities geologists use today to classify different materials. Although he also classified the stones according to how much of the elements of "earth" or "water" made up a type of stone, he also had the insight to record the reaction of different materials when they were exposed to fire. Theophrastus also carefully recorded where the stones were found and he was the first person we know of who cataloged a mineral that could be used as a fuel (most likely lignite, a vegetable derivative also known as brown coal).

Theophrastus was also the first to prepare several pigments that are know known to be dangerous, including white lead (lead carbonate) and cinnabar. He could classify alloys into their component parts and identify the materials of which they were composed. "On Stones" was one of the few truly scientific discussions of stones that existed in the world at that time.

He also took a similar path with two works in botany – "Causes of Plants" and "Inquiry Concerning Plants," which would remain authoritative resources for farmers and gardeners throughout the middle ages. Although these works were less centered around categorization than his geological text, he did apply the same scientific mindset to this field of study. Theophrastus carefully recorded the characteristics of many varieties of plants, shrubs and trees. He did believe that plants (especially small ones) could spring to life from inanimate material, which was a widely-held belief at that time. However, he also noted that if the theory, proposed by Anaxagoras, that seeds were carried by wind were true, than this path would be more likely to be the cause of most of the wild propagation of plants.

When Theophrastus died in 287 BCE, control of the school went to Strato of Lampsacus. Strato loved studying nature and especially the physics of moving bodies, a field that today would be known as dynamics. His ideas on gravity differed from those developed by Aristotle. Whereas Aristotle believed that bodies felt two forces,

---

religious minority which existed in pre-Islamic Egypt) and Ancient Greek. It would be the key to finally deciphering the language of Egyptian hieroglyphics. More on that will be coming up in the upcoming second part to this book series.

one upwards and one downwards, Strato held that only a downward force was required, since the falling object might be displacing other heavier material. This idea is central to the modern notion of buoyancy.

Strato wrote a book called "On Motion" in which he was able to deduce from everyday experience that objects fell faster towards the ground as they fell. One insight was that a long stream of water would be continuous near the top of its fall and broken up near the ground. This, he postulated, would not be possible unless the speed of the falling object increased over time. Also, he noted, an object dropped from a slight height would produce little impact, while the same object dropped from a great height would hit the ground with a bang. There was no explanation that Strato could find to account for this behavior, except that objects had to be traveling faster over time as they dropped. This basic truth of science discovered by Strato would have to wait 2,000 years to be re-discovered by Galileo at the turn of the seventeenth century.

Turning his insight towards the nature of air, Strato reinforced the earlier work of Empedocles by noting that an upturned jar inserted into water remained devoid of liquid, unless a hole were drilled in the bottom of the jar in question. Something, Strato realized, was holding the water away from entering the jar and he deduced, correctly, that the substance responsible was air.

Strato also designed a metal sphere with an airtight tube attached to it. Noticing that he could blow into the sphere without air escaping, he showed that air could be compressed, hence there had to be spaces between the particles that made up air. Second, he was able to suck air out of the metal ball without air entering into the system. Therefore, he knew that a vacuum could be produced, disproving Aristotle's assertion that nature abhorred a vacuum.

After the death of Alexander, the leader's kingdom (nearly the entire known world) was divided between his generals, with each general ruling his own territory as satrap, or governor. Alexander had given orders while on his deathbed that his empire would go "to the strongest", setting the stage for war.

One of his generals, Seleucus II (or Callinicus), was given much of the lands of the Middle East and Asia Minor, including all the lands which once made up the Persian Empire. This became part of the Seleucid Kingdom and it got off to a rough start. His brother Ptolemy III of Egypt invaded Syria and Mesopotamia, the Bactrians and Parthians successfully revolted in the east of his kingdom and Asia Minor was lost to his other brother, Antiochus Hierax.

Ptolemy I (also known as Ptolemy Soter, the Preserver or Savior[*]) ruled Egypt and Libya. He would go on to found a dynasty, as well as to help found the Great Library of Alexandria itself. He fought many wars and at one time, occupied Palestine. He also wrote of the conquests of Alexander and in 305 BCE, declared

[*] It is important to not confuse the different people from this era all named Ptolemy. The more famous Ptolemy, the astronomer, would not be born for nearly 425 years! And in the meantime, eleven other Egyptian kings would also be named Ptolemy.

himself pharaoh of Egypt. The dynasty he established there would last nearly three hundred years and end with the most famous (and final) pharaoh of them all, Cleopatra.

The Building of the Great Library 295 BCE

Ptolemy I, the man who guided the project to build the Great Library

Public domain photo

If there were an eighth wonder of the ancient world, few historians would doubt that the Great Library of Alexandria would be a superb choice for the honor. Its 500,000 to one million papyrus scrolls made the Library the greatest repository of knowledge up to that time. It was not the first library in Egypt – papyrus libraries had existed in that country since 3200 BCE, but this was far and away the largest, greatest library in the ancient world. More than just a library, though, this complex of buildings also contained massive lecture halls, accommodations for thousands of students, medical research halls, a zoo and far more.

Ptolemy I, a childhood friend of Alexander the Great (who always had a great respect and love for science and learning, having been a student of Aristotle) founded the library. The young leader assigned Demetrius Phalereus, an Athenian orator and former governor, to construct a great hall of learning in the new Greek colony of Alexandria. The idea for the library itself may have been suggested to Ptolemy by Phalereus as a way of gathering more information about military strategies and tactics.

Apart from Alexander the Great's conquest of Egypt, what were the reasons for the library being located in the Land of the Pharaohs? As usual, there are many factors that led to this shift. Although Egyptian studies of nature did not (for the most part) lead to innovation and discovery, they kept very careful records (you would too, if it were your job to keep re-drawing property lines along a river which flooded every year!). In fact, the ancient Greek word for Egypt is *khem*; this may be the root of our modern word chemistry.

The first library was small and would soon be outgrown. There were three main buildings within the complex by the time the main construction was completed under Ptolemy II. The main library was near the museum* (the building where classes were held) situated on the royal palace grounds overlooking the harbor. There was also the Serapeum, or Temple to Serapis, the Greek and Roman God of healing. This building held several thousand scrolls, as well as accommodations for up to fifty student/scientists at a time. It was designed in a similar fashion to the Great Hall.

In total, there were ten halls, each dedicated to one of ten branches of philosophy. The buildings each consisted of a main reading room, with separate

---

* The word museum derives from the Muses, ancient mythological creatures who trapped sailors with their enchanting songs.

smaller rooms surrounding each one, reserved for lone students or small groups researching a subject. Within each of these halls lay tens of thousands of manuscripts, collected from around the world and throughout hundreds of years of history.

The collected works were in the form of scrolls at first, the best of these wrapped in a fine leather casing. However, after the Roman takeover of Alexandria, they would be found in book form and stored in wooden cases called *armaria.* The shelves were numbered and at least after the Romans took control, they were fairly well maintained.

However, works on science, art and philosophy were not the only subjects that filled these shelves. The directors and the people who worked at the library also collected great works of fiction, plays and epic dramas. Physical researchers mingled with historians, theologians, poets, novelists and other intellectuals who wished to be surrounded by the greatest minds of their day. The idea of art and beauty as an integral piece of learning about nature had remained. The city became known as "The pearl of the Mediterranean."

The museum connected to the Great Hall through a magnificent hallway made of the same white marble and stone that made up the Great Hall and Museum. These buildings were architecturally harmonious and fit together perfectly. The Egyptians had learned extensive building and engineering skills from the construction of pyramids and other tombs centuries before. These included smoothly rounded rooms and doorways, styles of which may have found their way into the construction of these buildings.

The main hall was located in the northeast section of the city, near the harbor and the palace grounds. The land had courtyards, gardens and a zoological park. The park held exotic animals (and presumably plants) from around the eastern hemisphere. There was little need for legends of exotic animals here, as they could be seen on a stroll across the grounds any day. In the center of the Great Hall was a rounded dining room and above that was a domed observatory, surrounded by classrooms. People pondering the world or just taking a break from a class may have brought food to the observatory to eat during the day as they read the latest theories of other philosophers.

Within fifty years, the library was filled to capacity and it was decided to build a daughter library to house the surplus works. This was added to the Serapeum, housed south of the city, in the Egyptian quarter. The Serapeum is all which is left today of the ancient library and is just a stone's throw away from where the new Library of Alexandria now stands.

It was at this family of buildings that knowledge would become an asset in itself, where gold was less valuable than the written word. The chief librarian was selected from among the greatest scholars and teachers of the day and was considered such an important post, that the position was decreed by the king himself. With this support came the drive to aggrandize the library with rare manuscripts and the greatest minds of the day. In return for working at Alexandria, the philosopher/scientists here were easily able to making a living doing science alone, unlike any other time and place in the ancient world.

Ships coming into the harbor of Alexandria (then the largest in the world) were regularly boarded and searched. They were not being searched for contraband, but were instead searched for books. Any scrolls found on the ships would be confiscated and brought to the library where they were kept and copied. The library would then fill their shelves with the copies and the original documents were then returned to the owners*.

The library assistants who were responsible for the maintenance and organization of the scrolls and books were called *Hyperetae*. They were also accompanied by scribes who would edit the books, being paid according to the length of the works they transcribed, with additional bonuses going to the best scribes in the library. Some of these books were on a single subject and they were known as *amigei*, the multi-subjects books were called *symmigeis*. The collected works were listed in tables known as *pinakes*, which listed the books according to subject and author, much like the old-fashioned card catalogs that existed before the advent of library computer systems. The pinakes also listed a critical review of each of the authors in questions to further aid readers who wished to find a work on a particular subject.

This was when Greek culture was thriving from the prizes and riches brought from the Middle East (especially Persia) by the conquests of Alexander the Great. The Ptolemys used the lands they controlled to bring in more funds for all of their governmental projects, including the Great Library. This money was used not only to purchase new materials for the library, but also for stipends and grants for the chief librarian and researchers in Alexandria. Because of this great wealth, the directors of the library could often offer more money for available texts than any other library at the time could afford to pay, including the library in Pergamum. Among the early acquisitions of the library was the book collection of Aristotle. These were later sent to Rome.

The city of Samos gave rise to the philosopher Epicurus in 341 BCE, where he served in the Athenian military and later founded his own school, which he taught in the garden of his home. He taught that pleasure was the supreme and ultimate goal in life and that one should attempt to conquer fear of life and death. It is from this teacher that we get our modern word *Epicureanism*. With the rise of the philosophy of Epicureanism, the two great schools of philosophy, Stoicism and Epicureanism, began to compete for the dominance of the minds of the ancient world. One important difference in the physical models of nature that the two schools developed was that the Epicureans believed in the idea of "atoms and the void" developed by Democritus, with the differences we see in materials being due to the arrangement of atoms. Stoics, on the other hand, believed that objects moved through an infinitely divisible medium like a fish through water and that there were qualitative differences between the atoms

* Although it is said that, on occasion, the library would keep the originals and return copies.

of differing materials.

The Stoics held on to the belief that the universe was a living being, made of interconnected parts, having a mind of its own, purposely designed and riddled with fate. The Epicureans chose to believe that the atoms behaved according to laws of nature to produce the world around us. Each of these ideas were based on mere speculation. Most members of both groups rejected hard evidence and the use of mathematics to uncover underlying truths through hard numeric proof. However, each did stumble upon part of what most scientists know believe in our modern notion of physics. The atoms are separate from one another, with a void between them. However, the electromagnetic forces between them do cause interactions, from magnets to stubbing our toe on the bed stand.

Epicurus, of course, believed in atoms like Democritus and he was correct that some of the objects we see around us are elements, consisting of one type of atom, while others are compounds of various different elements. He also foretold the concept of natural selection, believing that nature shapes the organisms in a given area and that better-adapted members of a given species propagate to a greater degree than less well-suited members of a species. However, he also wrote on more than one occasion how the study of natural phenomenon should only be pursued as long as it leads to greater serenity and that there were boundaries of knowledge that the inquiries of science should never cross. Epicurus died in 270 BCE, in Athens, after living his last thirty-six years in the city.

Around that time a woman in Athens by the name of Agnodice became known as a great physician, despite the fact that women were not allowed to practice medicine at that time and place. It is uncertain to this day whether Agnodice was a real person, or if she were just a legend, but if the story is true, she was one of the early pioneers of female liberation.

The legend tells that Agnodice was born sometime around the year 300 BCE in Athens and she grew up wishing to study medicine. Since this was forbidden, she cut her hair and dressed as a man, attempting to learn what was known at the time about medicine. She studied under the physician Herophilus in Alexandria, learning her chosen trade.

One day, after her studies were complete, she heard the sounds of a woman in labor and rushed to help her. Believing she was a man, the woman refused her help. Agnodice lifted her tunic to reveal that she was, in fact, a woman. Soon, other women in the area, who were weary of male doctors, began to get treated by Agnodice. The other doctors, believing she was a man who was getting an unusual number of female patients, accused Agnodice of seducing "his" female patients (who they claimed faked illnesses in order to be treated by Agnodice) and brought "him" to trial.

At the trial, Agnodice proved to the court (in the same manner she had shown the birthing woman) that she was indeed a woman. At first, this was a terrible mistake, for the doctors present (all male) got even more upset that a woman should be practicing medicine. Their wives then arrived in the court and condemned their spouses as enemies for bringing to trial the person who brought healthcare to them for the first time in history. Athens then amended their laws so that any freeborn woman could practice medicine in the city and the first female physician we know of in Athens (assuming she was a real person) was free to care for her patients.

Meanwhile, at the library, the people there were about to greet a man so knowledgeable that his name would remain well known even to our present day as possibly the greatest mathematician who ever lived.

The observatory above the dining room in the Great Hall as it may have appeared at the beginning of the first century BCE. Although this room later had a dome, at this time the roof was likely open for observation of the heavens. This rendition is based on ancient accounts and shows a more Egyptian style of building than is shown in other portrayals of the Library.

Drawing by, and courtesy of, Leona Swett

Euclid c. 320 - 265 BCE

Euclid of Alexandria

Photo believed to be Public domain

Euclid lived and taught in Alexandria at the same time as Ptolemy I was still wandering the halls of the Great Library. The city of Alexandria, with a population of 300,000 people, had quickly become home to the center of science in the ancient world. It was also the world's largest seaport and a great home to commerce.

No one knows for sure exactly where or when Euclid was born, possibly Megara or Tyre. It is not even known whether he was one person, or a committee of scholars at the new Alexandrian Library. There had been a philosopher from Megara, also named Euclid, who lived one hundred years earlier. If the works of Euclid were designed by a committee, it is possible they took their name from this earlier philosopher. Nevertheless, where or whenever, or even if, Euclid was born, the works left behind by Euclid are easily the greatest and most influential works on mathematics ever.

Drawing on earlier sources such as Eudoxus and Theaetetus*, his most famous work, "Elements," used a stylized format for teaching mathematics. He began the book with postulates such as "the shortest distance between two points is a straight line" and logically inferred seemingly difficult concepts from these basic building blocks. He also did early work on conic sections, possibly with the assistance of Aristaeus the Elder, another brilliant mathematician, just slightly younger than Euclid.

Although some of the postulates in "Elements" are not true in all cases and many would go on to criticize this lacking, no other person has reached the type of inclusive command of mathematics of this book, or the influence of these early works. Euclid had likely studied in Athens under the tutelage of some of Plato's former students. This background is likely what made him a Platonist, believing in the Platonic solids as the basis for a universal truth.

Several of the contemporaries of Euclid and Ptolemy I have stated that the political leader asked Euclid, after the publication of "Elements," if there were not an easier route to learn mathematics. The brilliant mathematician answered Ptolemy that "there is no royal road to geometry." One day while a student was learning mathematics he asked of Euclid, "What shall I get by learning these things?" Euclid called in his slave and had the slave give the student a few small coins, replying that the student needed to be given the coins, since he "must make gain out of what he

---

* Theaetetus (417-369 BCE) lived in Athens and greatly studied two of the five regular solids – the octahedron (made from 8 triangles) and the dodecahedron (constructed of 12 pentagons). His work contributed greatly to books X and XII of *Elements*.

learns."[1]

A few decades before the birth of Euclid and the construction of the Great Library, a man named Autolycus was born in the city of Pitane, which was one of the twelve Hellenistic cities in western Asia Minor. Autolycus worked in Athens, developing what is believed to be the first work on mathematical astronomy in his book "On the Moving Sphere." He also wrote "On Risings and Settings," a book about observational astronomy. He attempted to explain the variable brightness of Mercury and Venus using Eudoxus' model of interconnected spheres, but was unable to do so. However, now the field of astronomy was subject to investigation through the extensive use of mathematical calculations. Euclid would hold Autolycus in the highest regard for these important contributions to astronomy and mathematics.

Another hotbed for scientific research and thought in Euclid's day was Pergamum, on the Caicus River. This city was founded by Attalus I and expanded by his son, Eumenes II who ruled Pergamum from approximately 225 – 160 BCE. This library was founded primarily as a center for opposition to the research being conducted at Alexandria. It retained its splendor and importance through the Byzantine period along with the famous Acropolis in the lower part of that city that rose 1300 feet above the sea.

Meanwhile, in Rome, the Republic was fighting a war with Tarentum, who requested aid from the king of Epirus in the year 280 BCE. The king sent over 25,000 men and twenty elephants to help fight the Romans. The armies met first at Heraclea and later at Asculum. The Romans were defeated at both battles, but at a terrible cost in lives to the army of Epirus. After the battle of Asculum, the king of Epirus stated "One more victory like that and I'll be ruined."[2] It is from the name of the king of Epirus, Pyrrhus, that we get our modern phrase a *Pyrrhic victory*. The gladiatorial games, which were once a province of funerals in Rome, also become a spectator sport at this time, creating a defining characteristic of Rome in our own time.

That same year, in Rhodes, a sculptor by the name of Chares completed construction of a bronze statue of the Sun God Helios that overlooked the harbor. It stood thirty-five meters (120 feet tall) and took twelve years to complete. This statue became known as the Colossus of Rhodes and was one of the seven wonders of the ancient world.

Although it is his most famous series of books, Euclid wrote books other than Elements. He also wrote a book entitled "Data," in which he gives ninety-four propositions of how the properties of forms can be calculated from other data about the object. He also wrote "On Divisions," which shows how various shaped objects can be divided into shapes of equal area to each other. "Phaenomena" is a work on

---

[1] University of St. Andrews. "Euclid." <http://www-groups.dcs.st-and.ac.uk/~history/Mathematicians/Euclid.html>
[2] Philip Matyszak. "Chronicle of the Roman Republic." 2003.

astronomy, containing tables that showed the rising and setting times of several stars. "Optics" was a work on perspective and he wrote four books on conic sections. Perhaps the most intriguing book by this remarkable mathematician was "The Book of Fallacies." In this book, Euclid talked about how scientific jargon can be used to dupe the unsuspecting into false conclusions. He had realized the danger of faulty science used for personal gain and tried to show people how to avoid such deceptions.

Near the time of the building of the Great Library, scientists began to disassociate themselves from the study of philosophy, studying hard science exclusively. Dissention was normal and accepted in Egypt and science flourished. The country of Egypt contained many years worth of astronomical records and with those, combined with foreign research they were able to obtain, Greek astronomy reached its greatest point.

The Greek scientists in Alexandria (notably Herophilus and Erasistratus) also began the first systemic study of the authentic, sometimes living, human body. For the first time in history, the doctors here began to cut apart human bodies, often in dissections of corpses, other times in vivisections of live prisoners donated to medical research. The medical community greatly benefited from the information gathered by the study of organs while they were still functioning within the body, however, the practice also generated a lot of opposition. Supporters of this practice held that such measures were understandable and worthwhile if they benefited the greater good. This debate would rage for centuries.

Herophilus had been born about the year 335 BCE in Chalcedon (modern day Kadikoy, Turkey). Through his studies, he became known as the Father of Anatomy, correctly identifying the many of the facts behind the workings of the liver, pancreas, eye, reproductive and the salivary organs. He was also the first to identify the function of the brain as the center of the nervous system and the role that arteries play in the circulation of blood throughout the human body. His studies were not flawless, however, as he also maintained that the optics nerves were hollow, which they are not. However, in his dissection of the eye, his description of the part of the eye that received the incoming light as "net-like" (*retiform* in Ancient Greek) gave us our modern term "retina" for this same component.

Another contribution to the modern day developed by Herophilus was his use of the measurement of a person's pulse as an indicator of disease. His teacher, Praxagoras, had talked earlier of the use of measuring a person's pulse, but he thought that the beating was an innate feature of arteries. It was Herophilus who realized the heart was responsible for the pulse in arteries as it pushed blood throughout the circulatory system. He also coined the terms *pararrhythimia* for a pulse with two distinct overlying beats, as well as the names of *heterorrhythnia* and *ekrhythmia* for two other types of abnormal pulse. The relationship between beats of a heart and music were not lost of Herophilus. The Greek musicians at the time refered to the upwards and downwards beats in music by the terms *arsis* and *thesis*. Herophilus chose these same words in his description of the two-beat rhythm of the human heart.

Herophilus wrote books on diet and pharmacology and a few of the actual prescriptions he wrote are still known. He also wrote an extensive treatise on anatomy, as well as a book about Hippocrates. Both of these works were lost after his death in 280BCE, at the age of fifty-five.

About five years before the death of Herophilus, a young new doctor opened a school at Alexandria. His name was Erasistratus and despite just reaching adulthood,, he had already been court physician to the king of Syria. He quickly joined Herophilus in the study of living and deceased humans despite the rancor that such research incurred.

Whereas Herophilus had identified the brain as the center of the nervous system, it was Erasistratus who traced out the nerves of the human body as they connected to their central network. He also developed a map of the arterial and venous systems, including how these channels connected to and from the heart, including realizing the proper function of the four main valves of the heart as one-way valves. He also correctly deduced that food was carried through the digestive system through the contraction of muscles.

His notions of the functions of these networks were accurate in some ways and erroneous in other ways. He defined the signals carried by the nerves as "nervous spirit" and he believed the arteries carried "animal spirit", yet he also realized that the heart was responsible for distributing air from the lungs throughout the body. However, he believed that the arteries were filled with only air, distributing the gas to each different part of the body, rather than the correct notion of oxygen-rich blood. This was a simple mistake for him to make, since the term *arteria* meant both arteries and the tubes of the respiratory tract, including the windpipe. The Ancient Greek term for the windpipe in humans was *he tracheia arteria*, or "the rough artery." It is from this phrase that we derive our modern word *trachea*.

Euclid died in the year 265 BCE, fifteen years after Herophilus passed. Erasistratus died fifteen years after that, in the year 250 BCE, aged fifty-four. The very same year that Erasistratus passed away, the Great Library also lost the greatest astronomer of the age: Aristarchus of Samos. However, before his death, Aristarchus would realize, almost 1,800 years before Copernicus, that the Earth went around the Sun, making the Earth just another planet.

Aristarchus of Samos c. 310 - 250 BCE

Aristarchus of Samos

Photo believed to be public domain

Perhaps two years before the birth of Aristarchus of Samos, Rome constructed the first of what would become nine aqueducts, the Aqua Appia*, that stretched for 16 km (10 miles) underground. At the height of the aqueduct system in Rome, the city was supplied with thirty-eight million gallons of water a day. This is about one percent of the capacity of the modern Los Angeles aqueduct, which supplies that city with four billion gallons of water daily.

Aristarchus was born in the city of Samos and studied under Strato of Lampsacus in Alexandria, probably while Aristarchus was in his mid-twenties.

About the same time, in 285 BCE, the king of Egypt and founder of the Great Library, Ptolemy I Soter, abdicated his throne to his son, Ptolemy II Philadelphus ("Ptolemy the Brotherly"). This second Ptolemy quickly made Egypt the dominant maritime force in the Mediterranean and brought the country's economy under government control. He spent a great deal of the money of his government on filling the library with books from around the world, greatly expanding the number of works upon its shelves.

Ptolemy the Second named Apollonius of Rhodes, a poet who excelled in lengthy epic poems and the author of perhaps 800 books, as chief librarian of the Alexandrian Library. Some people have books written about them. Apollonius had a book written about him 450 years after he lived. Philostratus, near the beginning of the third century CE, wrote the work, "Life of Apollonius." In the book, Philostratus paints Apollonius as a super-human, wise and knowing.

Born in the city of Tyana (now Bor, in southern Turkey), Apollonius grew to be a member of the neo-Pythagorean school of philosophy, as well as director of the Great Library of Alexandria. He worked to reform cult society in his homeland and traveled extensively, meeting several Roman emperors.

The only work of Aristarchus that survives today is entitled "On the Dimensions and Distances of the Sun and Moon." In this work, he used a reasonable method to determine the relative distances and sizes of the Sun and Moon, but his mathematics was lacking and the instruments of the time were too primitive for an accurate reckoning. His final answer of the Sun being twenty times further away from the Earth as the Moon is only five percent of the correct answer, but if his initial

* The aqueduct was named for Rome's leader at the time, Appius Claudius Caecus, for whom the Appian Way is also named.

assumption of one of the angles in the Earth/Sun/Moon triangle had been correct, his method would have come close to divining the correct answer. Also, in this book, he realizes that the Sun and Moon subtend the same angle (otherwise total eclipses would never occur),* however, he gives a figure of two degrees for this angle. This is four times larger than the correct answer. However, in later works, he is said to have used the correct figure of one-half of a degree for those bodies as seen from Earth. It is possible that the lone surviving work was one of his earlier papers and that he learnt the correct figure later.

A coin from the reign of Ptolemy II. On the face is Zeus and on the back is a spread-winged eagle on a thunderbolt.

Courtesy Ancient Coins Canada

Aristarchus also had another remarkable insight – he realized that the complicated network of celestial spheres postulated by most scientists of the time, including Eudoxus, could be greatly simplified by simply moving the Earth from the center of the planetary system and, instead, having it revolve around the Sun. This was an early heliocentric model, with the Earth revolving on its axis once every twenty-four hours. The celestial model of Eudoxus required eight spheres to explain the motions of the daily movement of the heavens across the sky – Aristarchus did the same with one motion of the Earth.

The heliocentric theory of Aristarchus created a prediction that made many people reject his notion. For, if the Earth were revolving around the Sun, then the stars on the outermost sphere would seem to slightly shift their positions as the Earth traveled from one side of its orbit to the other. This was not observed, so the theory never gained popularity. However, the instruments at the time were too inaccurate to measure the small shift in parallax that occurs when looking at even the nearest stars. This small shift in the apparent positions of the stars was not actually measured until the late 1830's, by F.W. Bessel. The people who favored a geocentric universe also realized that a parallax should be seen when looking at the stars from various places on the Earth. However, this could be explained by reasoning that the diameter of the Earth was incomparably small in relation to the distance to the stars. For the heliocentric model to be correct, than the diameter of the orbit of the Earth around the Sun would have to be much smaller than the distance to the stars. Although Aristarchus was correct in this notion, such distances were unfathomable to most of the people of ancient Greece. The theory of Aristarchus was better than the engineering of the day and the theory lost.

---

* You're here at a lucky time on Earth. Due to the bulge created in the Earth by the gravitational pull of the Moon (creating tides on Earth) and the unequal difference in rotational speed of the Earth and Moon, the Moon is slowly moving away from the Earth and total solar eclipses will one day be a thing of the past.

Many Greeks also rejected the heliocentric model based on religious and spiritual beliefs in the sacredness of the Earth. Having our home world traveling around the Sun made the Earth just another planet. This was an idea that was distasteful to many people of the time and there was talk of imprisoning Aristarchus for impiety for this belief. However, many of the scientist/philosophers of the era also rejected the heliocentric model for more logical reasons, even if they were wrong in their objections. Aristotle had taught that heavy objects fell towards the Earth. Naturally, they reasoned, the Earth is already where it would tend to go towards, therefore it could not move. It was also believed by many that any object traveling through the air could never overcome the speed of the Earth's rotation, therefore they figured, nothing could travel eastwards, or it would be overtaken by the ground rushing up behind it.

Aristarchus is likely one of the least appreciated of the ancient philosopher/scientists, but he was ingenious at combining mathematics with astronomy, even if his astronomical observations could sometimes suffer from his great attention to mathematics. Perhaps the greatest contribution he made was being able to show that through mathematical means, one could truly measure the immeasurable.

Archimedes 287 - 212 BCE

Archimedes of Syracuse

Photo believed to be public domain

Archimedes was born in the year 287 BCE, in the city of Syracuse, Sicily, the son of an astronomer named Phidias. He would go on to become a master of machines, gears and levers. It is he who is famous for the words, quoted by Pappus of Alexandria in the mid-fourth century CE, "Give me somewhere to stand and I will move the earth." Archimedes was a man of many interests, doing work in physics, astronomy, arithmetic (including geometry) and the design of war machines.

His birthplace of Syracuse is traditionally accepted to have been founded in the year 734 BCE, by Greek colonists from Corinth. For 450 years, Syracuse was an independent, proud, city-state on the coast of Sicily. Two years before Archimedes was born, the tyrant Hiketas gained control of Corinth and his despotic rule would continue until 279 BCE.

Perhaps fueling childhood love of science and engineering in Archimedes were stories told to him of the building, at that time, of the great Lighthouse at Alexandria in 280 BCE. Lighthouses had been built in the Aegean since the seventh century BCE, however, when this project was completed, it was the most powerful lighthouse in the ancient world. The light from the Pharos Lighthouse at Alexandria, reflected by mirrors, could almost certainly have been seen up to thirty miles out to sea. Some even claimed that the light was so powerful, it could even set boats afire from that distance, although this seems unlikely.

The fire at the top of the lighthouse was fueled by an unknown source; wood was (and continues to be) rare in Egypt, so this great lighthouse may have run on animal dung. The lighthouse stood until the fourteenth century CE, when it was destroyed in a series of earthquakes. The materials from the great Pharos Lighthouse may now be a part of a fort, which guards the port of Alexandria to this day.

Archimedes was in his early twenties when the first of the three Punic Wars between Rome and Carthage erupted in Sicily in 264 BCE. Legend tells that Rome and Carthage were destined for war from long before the time when Rome was founded. The story known to every Roman child was of Aeneas, said to be the father of the Romans, who four hundred years before the founding of Rome, was carried from the city of Troy as it was being sacked. He traveled for a while, including visiting Carthage and he finally left for Italy, where Aeneas married an Italian princess. However, while he had been traveling in Carthage, he is said to have fallen in love with the Carthaginian queen, named Dido. He left Dido, who became so distraught that she committed suicide. The son of Aeneas, Ascanius, founded the city of Alba Longa, which is where Romulus and Remus were said to have come from before they founded the city of Rome. Thus is the legend of the bad blood between the two ancient cities, but the real reason for the Punic Wars had much more to do with economics and control of the Mediterranean.

The Carthaginians ruled western Sicily and the Romans had just gained control of the southern tip of Italy. When Messana, on Sicily, rose up against Carthaginian control, Rome entered the fight against Carthage and the First Punic War ensued. Carthage was originally settled by the Phoenicians, and the people of city were known to the Romans as the Phoenicians, or *Poeni*. It is from the word *Poeni* that the Punic Wars derive their name.

The impetus of these wars would drive Rome to build their first great navy and defeat the Carthaginians. The Romans were not naturally sailors and when Rome attempted to form its first real navy during the First Punic War, they would find that their inexperience and the weather would sink more ships than the Carthaginians. The First Punic War began in the waters surrounding Sicily in 264 BCE and was the first and last war fought by Rome almost entirely at sea. In fact, as the war raged on for years around the island, there is not a single battle known from that war which occurred in the inland of Sicily. It took years before the Romans were a match for the navy of Carthage, but once the Romans took dominance of the seas, they would remain superior for the rest of the Punic Wars.

So what gave the Romans the leap in technology they needed to modernize their navy? Early in the war, Carthaginian ships were harassing Roman ships as they passed through the Mediterranean. One of these ships caught on the shore of a small island and the Romans captured the prize. They then used this as a model to build a new fleet of ships based on the same design. This fleet of ships was headed by Gnaeus Cornelius Scipio, who would become consul in 260 BCE. As the fleet was being built, Cornelius Scipio trained his new navy how to row by having them sit on beaches, set upon beaches in front of where the ships were being constructed, and practice rowing in the air.

When Scipio heard that the Carthaginian-allied city of Lipara was lightly defended, he took seventeen of his new ships to the city to win her to the Roman side. When he got to the island that contained Lipara, he did not realize that 150 Carthaginian ships were nearby, ready to pounce. The leader of the Carthaginian fleet was named Hannibal (no relation to the infamous Hannibal that will come into the story soon) and he sent seventeen ships to head off Scipio. The Carthaginian ships arrived in the middle of the night, blocking Scipio and his men in a harbor. The next morning, realizing they had been trapped, the Romans (including Scipio) panicked, beaching their ships and abandoning the fight. Many of the Roman sailors were able to escape onto the island, but Scipio and a good portion of his men became captives of Hannibal. Because of this act, Scipio was given the nickname *Asina* – a derogatory name, meaning a woman's posterior.

Hannibal could have stopped here, but he decided instead to find out where the rest of the Roman fleet was hiding. He found the rest of the fleet alright and when he did, the Romans destroyed nearly the entire Carthaginian fleet and almost killed Hannibal.

In 249 BCE, a Roman fleet was led by Publius Appius Claudius Pulcher, one of the consuls for that year. Before heading into a sea battle with the Carthaginians, Pulcher decided to consult with the Gods to see if he would be victorious in battle. The method he chose to do so was to follow a tradition in the Roman navy of spreading out food for chickens that were on the boat and seeing whether or not they would eat. The legend said that if the chickens ate heartily, the battle would be a success. Sure enough, the chickens would not touch the food laid before them. Rather than take this as a sign of foreboding, Pulcher instead decided that if the birds would not eat, they would drink instead and he had them thrown overboard.

The Romans had meant to attack the Carthaginian fleet in a harbor, but by the time they got there, they found that the Carthaginian fleet, led by Adherbal, had already left the harbor. The Roman ships attempting to leave the harbor became entangled with the ships still entering the harbor and collisions and mayhem ensued. Adherbal and the Carthaginians came back around the trapped Roman fleet, sinking ninety-three out of the one hundred and twenty ships led by Pulcher. He was later tried for incompetence, but no mention was ever made of the superior military know-how of the drowned chickens. Two years later, the sister of Pulcher, Claudia, was stuck in heavy foot traffic at a festival and remarked that it was a shame her brother was not still around to lose another fleet, since in her opinion, the Roman populous was in need of further culling. She was heavily fined for making this statement.

The result of the First Punic War would be the seceding of Sicily to Rome and the further annexing of Sardinia and Corsica by Rome in 237 BCE. However, Rome did not hold on to this power once they had achieved it. After the First Punic War, they left control of the Mediterranean to other lands such as Rhodes. The Romans did not wish to take over the responsibility of policing the Mediterranean for pirates, and they also did not see the major tactical advantage they were abandoning. This was until Rhodes later provoked Rome, and they lost their fleet in the aftermath. The pirates now ran rampant throughout the Mediterranean. The result of the flourishing of piracy was that Rome had to rebuilt her navy and take control, once more, of the sea.

Archimedes is probably most well known for his invention of the Archimedes' screw. This is a device that is used to remove water from one area to move it to another place. It has a helix shaped[*] inner rod inside of a tube and one side of the tube is placed in the water. As the inner screw is turned, the water is forced up through the ridges and comes out the top. The Greeks also found that the Archimedes' Screw could be used as a new type of wine press. This irrigation device is still in use today in many parts of the world and was used extensively in the "reclaiming" of land from the

[*] A helix is similar to a screw with extremely deep ridges.

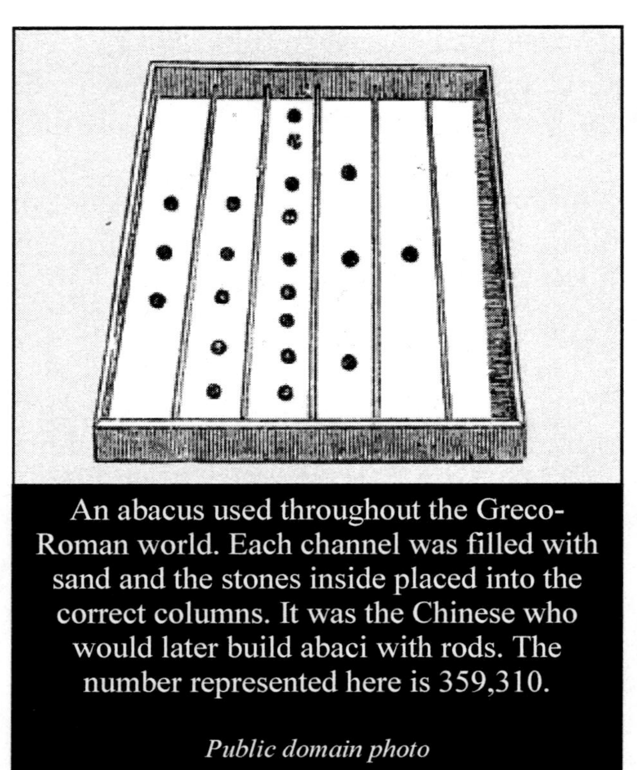

An abacus used throughout the Greco-Roman world. Each channel was filled with sand and the stones inside placed into the correct columns. It was the Chinese who would later build abaci with rods. The number represented here is 359,310.

Public domain photo

ocean to form the modern country of the Netherlands, forty-two percent of which was once under water.

Archimedes would write to his friends in Alexandria about his new theories and discoveries, without the final proof so that they could figure it out for themselves and often enclosed personal notes with the scientific and mathematical works. However, some unscrupulous people within the library began to copy his work and claim it for their own. Therefore, the last time Archimedes sent work of his to the Great Library, he also sent along two theorems that were incorrect. This was so that anyone who claimed these theorems as their own would be exposed as frauds.

One of the greatest insights into mathematics that Archimedes gathered concerned the area of the curve in a graph. The method we use today to figure such numbers from a given set of data is known as integral calculus*.

Although the system Archimedes used for this, known as "the method of exhaustion" was not perfect, he was on the verge of an evolution of a mathematical technique that would give him tools which would not be discovered until the early eighteenth century by Leibniz and Newton. Archimedes is also known to have found methods to fairly accurately estimate square roots of large numbers and to have developed a fairly good estimate of the value of pi.

Archimedes was so devoted to his mathematical work (especially geometry) that he would rarely even take baths, since they took time away from his calculations. At times, his friends would physically grab Archimedes, undress him and force him into the bath. When they did so, however, Archimedes let it interrupt his work as little as possible, drawing figures and doing geometry on the side of the tub, or upon his body, using his fingers and the oil which was set there for anointing.

One problem, however, stumped Archimedes for a while - how to measure the volume of irregular shapes. Hiero, the king of Syracuse, had commissioned a new

* One simple example of integral calculus is the relationship between acceleration, velocity and distance. If one were to plot the changing acceleration of an object (say, a moving car) on the Y (vertical) axis of a graph and time on the X (horizontal) axis, the area under the curve is equal to its average velocity during that time. If you plot the changing velocity of the object vs. time, the area under the curve is the total distance traveled.

crown to be made by a local metalworker. The metal worker received a good amount of gold from which to make the crown and he returned a golden crown that had the same weight as the original mass of gold that he had been given.

But the question occurred to Hiero – Was the crown entirely gold, or had the metalworker diluted the gold with another alloy and pocketed the remaining gold for himself? If the exact volume of the crown were known in addition to its mass, then the density of the object could easily be figured. If the density of the crown were different than the density of pure gold (which was known), then the metalworker stole some of the gold he was originally given and substituted a less valuable metal, cheating the king. However, no way was known of measuring the density of an object with such an unusual shape.

The answer, surprisingly enough, came to Archimedes in the bath.* As he sunk into the water, he noticed the level of the water rise. He may have slowly raised his body in and out of the water several times, noticing the water level going up and down as more or less of his body was under the surface of the water. Then, it struck him. No matter what the shape, whenever an object is put into water (assuming it does not itself absorb water), the water is displaced *by the exact same amount* as the volume of the object. The answer to how to measure the volume of the crown was solved and it was as easy as putting the object into water and measuring how much the water level within the enclosing container rose. In a moment of discovery still famous today, Archimedes rose from the bath and ran through the streets of Syracuse, still nude, shouting "Eureka!" ("I have found it!"). In fact, the metalworker HAD cheated Hiero and he was executed for his crime.

In fact, it was to his friend and relative King Hiero that Archimedes made the remarkable boast that if he had a place to stand, he could move even the Earth itself. An extraordinary claim requires extraordinary evidence and Hiero challenged Archimedes to prove his boast. Unable to move the Earth itself (if only for a lack of a place to stand), Archimedes chose they next best thing. He went to the shore, where Hiero had his large warships in dock. Attaching ropes to one of the largest ships in Hiero's fleet (fully loaded with a full crew and complement of cargo) and running the ropes through a series of compound pulleys, Archimedes stood on the shore, holding a rope. Without even a great effort, he moved the ship, weighing several tons, without breaking a sweat. Archimedes showed that gears could give man super-human strength.

As Archimedes entered his forties, Egypt entered the height of its power under the Ptolemaic Dynasty, as Ptolemy III Euergetes (The Benefactor) came to power in 246 BCE. He quickly established Egyptian maritime dominance in the Aegean and invaded Syria to avenge the murder of his sister and infant nephew. Syria, at the time, was part of the Seleucid Empire, which stretched from Macedonia to Pakistan and the

---

* There's no news as to whether Archimedes got in that particular bath voluntarily, or was forced in by the people who had to live around him.

nephew of Ptolemy Euergetes who had been murdered was heir to the throne.

This war had an interesting, if minor, repercussion to the modern day. The wife of Ptolemy III, Queen Berenices, made a vow to fulfill if her husband returned victorious from this war. She swore that she would donate a lock of her hair to the local temple upon the victorious return of Euergetes and when he returned after winning the war, she followed through on her promise. The morning after she donated the lock of hair, it disappeared from the temple. An astronomer from Samos working at the library by the name of Conon (who was friends with Archimedes) claimed that he could see the missing lock of hair in the night sky, situated between the constellations Virgo, Leo and Bootes. People began to call this constellation *Coma Berenices* ("The Hair of Berenices"), a name it is still known by today.

We know of several books that were written by Archimedes on a variety of subjects, including finding the center of balance of various shapes, hydrostatics (the study of fluids at rest and the pressure exerted upon submersed objects), geometry and other branches of physics and mathematics. "On Plane Equilibrium" is likely the oldest book we know of that Archimedes wrote. This work was an exploration of how to find the center of gravity of various geometric shapes such a parallelograms, trapezoids and parabolas. "On the Sphere and Cylinder" and "On Conoids and Spheroids" as well as "On Spirals" are works in geometry and in "On Floating Bodies" he laid groundwork in the science of hydrostatics.

There are at least two other remarkable books by Archimedes that we know of today. The first was entitled "Measurement of the Circle," in which he deduced that the true value of pi had to lie between 3 10/71 and 3 1/7. He did this by drawing a circle and fitting two 96 sided polygons around it - one that just barely fit inside the circle and the other which barely fit around it. Therefore, the true value of pi (which is irrational – it has no last digit and cannot be expressed as an exact fraction) was limited by the combined lengths of each of the ninety-six sides of both figures. People today may have trouble understanding why the Greeks had such trouble grasping the concept of an irrational number, but we must remember that Greek mathematics was only beginning to realize that there were any numbers at all which could not be expressed as a fraction of two other numbers. Also, nearly any mathematical premise was not considered to have really been proven unless the method of proof could be shown using a compass and ruler.

Archimedes' most famous work, however, was "The Sandreckoner." In a grandiose mental experiment, Archimedes made an estimate of how many grains of sand could fit into the entire known universe. Using Aristarchus' heliocentric system of the universe along with the measurements of the size of the Earth provided by Eratosthenes and other estimates provided by Eudoxus and his father Phidias, he calculated that such a universe would hold $8x10^{63}$ grains of sand (in modern notation; this is an eight with sixty three zeros after it). This was no small accomplishment, since the Greek numbering system could not even express numbers of this magnitude and Archimedes was forced to invent a whole new numbering system in order to express numbers of this size.

However, it was not discoveries, mathematical or otherwise, that Archimedes would become best known for in his lifetime. Nor was it his great contribution to irrigation (and wine) with the invention of the Archimedes' screw. It was the remarkable military devices that he designed and built for the defense of Syracuse from the invading Roman armies who had laid a siege to the city in 212 BCE.

Archimedes had been convinced by his friend and relative, King Hiero, to turn his attention from abstract ideas to the more practical purpose of saving the city from the invading Romans. Archimedes began thinking and Hiero commissioned the order to build these machines. Archimedes soon realized that theory alone was not the best way to engineer these devices, so he set upon testing different variations of his models to find the best possible designs to defend his home city.

He devised machines that bombarded the invading armies with countless spears and boulders, destroying the lines of the Romans and sending many fleeing. Even the sound of the mighty boulders breaking upon the bones and armor of the men in the ranks was terrifying to the legions sent upon Syracuse. Archimedes created a new catapult that could launch boulders four hundred yards into the oncoming Roman lines or ships. Although catapults had been in use since 397 BCE at the siege of Motya, the catapult of Archimedes was superior to anything designed before.

He is also said to have invented machines that would launch large pieces of timber from the walls of the city into the hulls of the Roman ships. Those that did not sink right away from taking on water were often scuttled by a large boulder, attached to the pole now within the ship that came down upon it with a thunderous blow. Other machines attributed to Archimedes (though less probable ones) were cranes that lifted ships out of the water and dropped them (sometimes after shaking the crew out) upon jagged rocks. Any person or ship so dropped would have been instantly shattered upon falling.

There are also stories that Archimedes designed mirrors that would reflect the light of the Sun, focusing it on the ships, causing them to burst into flame. This, however, is unlikely to be true. The mirrors at the time were not of high enough quality to reflect enough light to make it possible and the engineering was not precise enough to design the mirrors to focus the light perfectly at a single point from any real distance away.

Perhaps the most remarkable of the war machines of Archimedes was his steam cannon. This looked like a civil war cannon, with two noticeable exceptions. The first was two pieces of wood, one that ran down the length of the barrel, holding the ball in place and the second piece of wood sat at right angles to the first piece, held against the mouth of the cannon by two metal brackets. The second difference had to do with the propulsion method, since ancient Mediterranean people did not have gunpowder. There was a water tank and valve above the breech of the cannon and a fire underneath it. The fire would burn until the metal was red hot and then the valve was turned, allowing water into the breech. Pressure was built up behind the cannonball until the wood at the front of the cannon broke into two pieces, shooting the cannonball out with a tremendous force.

Unable to take the city by land due to not only the weapons of Archimedes' design, but to the fear that those designs inspired in the Roman army and navy, the Romans decided to lay siege to Syracuse instead. The Roman soldiers were terrified of the weapons of Archimedes and would often run away at the sight of a small piece of wood or string, believing that it was a booby-trap laid by the elderly scientist. Sometimes they would even run in terror seeing an old man in a crowd, fearing that it was Archimedes himself.

Despite the terrible resistance they met, the Romans finally captured the city in 212 BCE, during the Second Punic War. The exact circumstances surrounding the death of Archimedes are unclear, but it is apparent that some soldier in the Roman military happened upon Archimedes, either working on a problem oblivious to the invasion, or carrying his instruments with him, which the soldier mistook for gold. Either through frustration with Archimedes refusing to leave his mathematical works, or through greed wanting to take the gold that was never there, the soldier ran Archimedes through with a sword, killing him.

The most popular notion of what happened when Archimedes met his fate is depicted here in an engraving by Giovanni Mazzuchelli (1707-65)

Public domain photo

Amazingly, the works of this great man and even knowledge of his existence, nearly completely disappeared from the minds of the people of the ancient world, except for the scholars of Alexandria. In 75 BCE, when Cicero (an up and coming political pundit in Rome who would become Senator the following year) went to find the grave of Archimedes, he found it hidden by brush and overgrowth, neglected for centuries. He cleaned off his grave, clearing the collected vines, out of respect for the great man. Archimedes' tombstone had a sphere within a column inscribed upon it: Archimedes' discovery, written upon the stone, that if a sphere barely fits inside of a cylinder, then both the surface area and volume of the sphere are two-thirds that of the cylinder.

The brilliance of Archimedes would remain virtually unknown until the sixth century, when Eutocius published commentaries on his work and the works of Archimedes were re-discovered by the outside world.

Eratosthenes of Cyrene c. 276 - 196 BCE

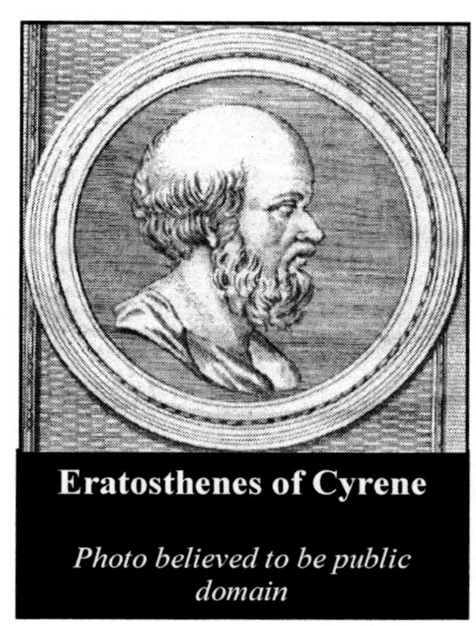

Eratosthenes of Cyrene

Photo believed to be public domain

Eratosthenes is most famous for a measurement that he made, for he was the first to measure the size of the Earth through direct means. He had taken a modest observation, performed a simple experiment and used basic mathematics to find the size of our home planet. Moreover, it is likely that he made this measurement with remarkable accuracy.

Born in Cyrene (which is now Shahhat, Libya), he spent his time studying a wide variety of subjects under numerous teachers, including the Greek poet Callimachus. After leaving Callimachus, Eratosthenes spent the next several years studying in Athens, where he furthered his esoteric knowledge of the world around him.

Eratosthenes was known as having a great deal of knowledge, in a great number of different fields of study, but not excelling to supremacy in any one of the fields. He was a sort of "Renaissance man" over 1500 years before the Renaissance, doing work in astronomy, mathematics and geography, along with the required philosophy, as well as history, poetry and doing work as a theatre critic in his spare time. For this, he became known by the name "Beta" (the second letter of the Greek alphabet) because it was told that he was "Second best in the world at everything."[1] He also earned (or was given) the name "Pentathlos," a name which was given to an athlete who was successful at several sports, but excelled at none. Eratosthenes would go on to become director of the Great Library for over four decades, from 245 BCE to just before the beginning of the second century BCE.

He had figured out how to calculate how much the orbit of the Earth around the Sun differs from a circle. Doing so, he found the correct answer within 1/10th of a degree. He also compiled a star catalog of 675 stars, but it was his measurement of the circumference of the Earth that he is best known for today.

How did he manage to do it? While working in the Great Library, Eratosthenes read in a scroll that in Syene (modern-day Aswan, Egypt, located on the Nile) on the longest day of the year (usually 21 June), vertical objects cast no shadow and that the Sun shone down in wells, being directly overhead. So, Eratosthenes, being the clever sort of fellow he was, performed an experiment. Come the solstice, he checked to see if vertical objects cast shadows near noon at Alexandria. Sure enough, they do. He knew the distance to Syene (possibly paying a runner to pace out the distance), which was about 800 km (500 miles) and he knew the angle the shadows of vertical objects cast in Alexandria on the solstice (about seven degrees). Since there are 360 degrees

---

[1] Sagan, Carl. "Cosmos." *The Shores of the Cosmic Ocean.* 1980.

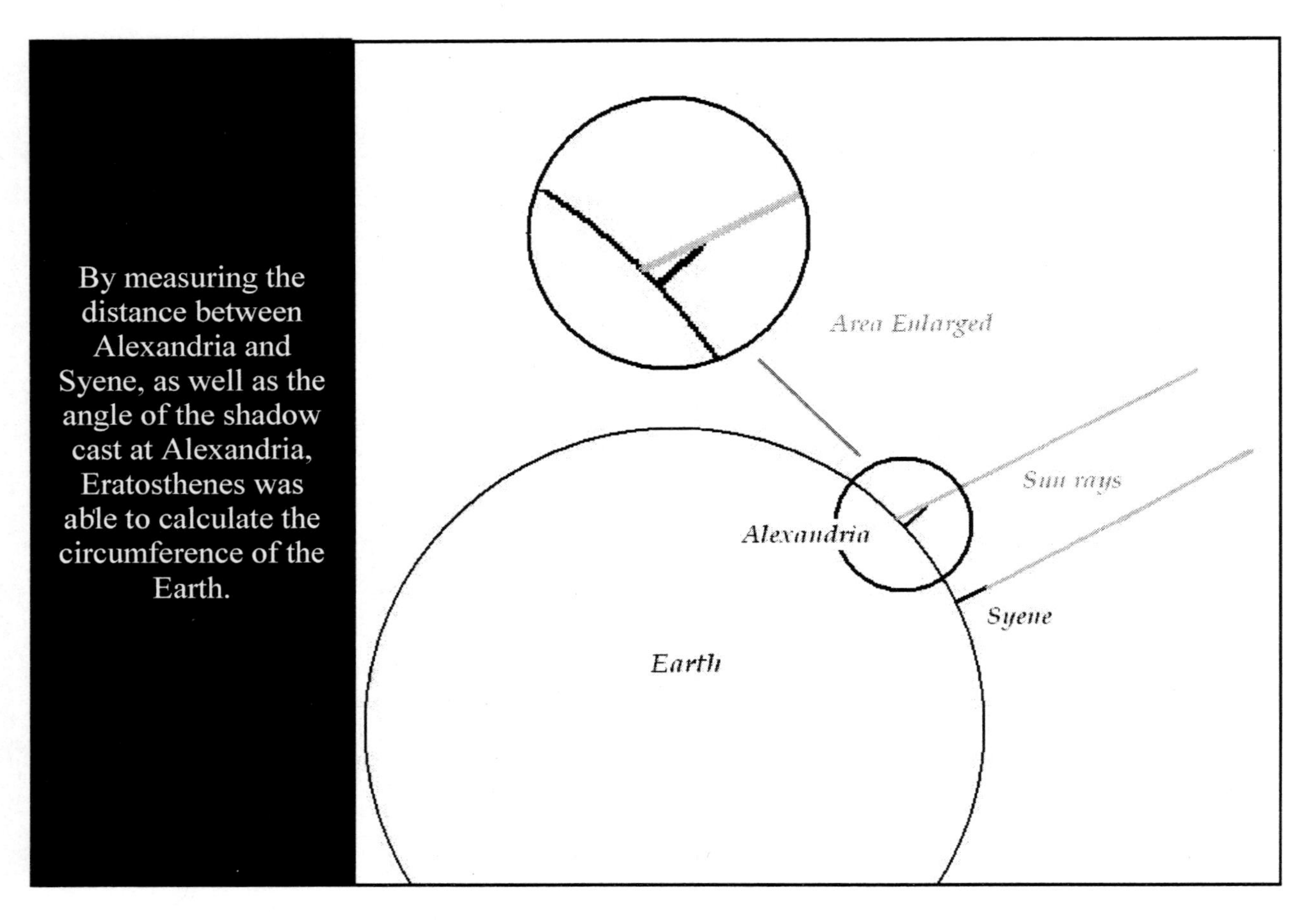

in a circle*, seven degrees is about 1/50th of that number. This means that the distance around the Earth had to be about (800*50) or 40,000 km (25,000 miles) around its perimeter. This is, within a few percent, the correct answer if he used one particular accepted length for the standard unit of measurement, the *stadia*.

He also measured the distances to the Sun and Moon, albeit less accurately. Using measurements he made during eclipses, he measured the distance to the Moon as about 125,000 km (78,000 miles) and to the Sun as less than 130 million km (over 80 million miles). Both answers are smaller than the correct numbers, but these conclusions show that he was thinking in the right scales.

Eratosthenes was nearly sixty years old when Hannibal made his infamous unsuccessful march to Rome, over the Alps, mounted upon dozens of elephants.

Hannibal had been born in 247 BCE to Hamilcar Barca, a Carthaginian general from the first Punic War. Barca had devoted his life to increasing and enhancing the power of Carthage in Spain, after their loss in the First Punic War. When Hannibal was only nine years old, before accompanying his father on an expedition to conquer Spain, Barca made him swear a lifelong hatred of Rome, the bitter rival of Carthage.

---

* Eratosthenes likely did not use degrees to make this measurement, since the modern notion of 360 degrees to a circle likely did not develop until Hipparchus, decades after the death of Eratosthenes. However, we can see, using degrees, the math that Eratosthenes used to make his measurement.

Hannibal grew up through his teen years at his father's side, amongst his soldiers.

In 229 BCE, when Hannibal was eighteen, Barca was killed and Hannibal was named the head of the Carthaginian cavalry in southern Spain. He joined his brother-in-law, Hasdrubal, in a seven-year drive to consolidate and reinforce the Carthaginian dominion over the Iberian Peninsula.

Upon the assassination of Hasdrubal in 221 BCE, as tensions continued to grow between Rome and Carthage, the position of Commander in Chief was bestowed upon Hannibal by the army. Quickly, much of what is now southern Spain was under his control, with the exception of one city loyal to Rome, Saguntum. Within two years, his army began a siege of Saguntum, which would last for eight months. Rome considered this a treaty violation and in 218 BCE, demanded that Carthage deliver Hannibal into their hands. They did not do so and this triggered the Second Punic War, which would rage for seventeen years.

Hannibal Crossing the Alps. From a fresco by Palazzo del Campidoglio; circa 1510. Currently in the Capitoline Museum in Rome.

Photo used under Creative Commons License.

Hannibal lead his 40,000 men and thirty-four elephants from New Carthage (modern day Cartagena, Spain) eastwards, towards the Pyrenees, with his sights on conquering the greatest superpower in the world, which had over 750,000 soldiers at her disposal.

He crossed the Pyrenees and the Alps and his elephants[*] crushed the Iberians. However, winter began to set in and took its toll on many of the poor lumbering beasts. They either froze to death from hypothermia, or developed such severe blisters on their feet that they could travel no further and were stopped in their tracks.

All told, Hannibal may have lost 15,000 men and the great majority of his elephants crossing the distance from Spain to northern Italy to cold, starvation, landslides and attacks from hostile tribes in the lands he and his men were crossing. However, he continued to get new men from among the friendly groups whose paths he crossed and essentially maintained the sheer numbers of his forces.

He then landed in northern Italy, where he continued fighting off Rome (even holding the upper hand militarily most of that time) and forging alliances with tribes

[*] In order to make the elephants more aggressive, Hannibal had fed them figs, which causes elephants to break out into an irritating rash, similar to eczema. He also plied his elephantine cavalry with alcohol for courage and aggression.

and city-states which had been friendly to Rome, attempting to not only weaken Rome with any switch in allegiances, but to strengthen his own hand at the same time. These alliances were exactly what had made Rome so powerful on the Italian peninsula and now Hannibal was undermining the power of Rome within Italy itself.

When the Spring of 218 BCE came around, Hannibal began to march down towards Rome. He defeated Rome at the battle of Ticinus and then at the battle of Trebia, where 20,000 Romans met their fates. The Romans were led by Scipio Africanus the Elder, whom Hannibal would see again in battle later in his life.

The next year gave Hannibal one of his most stunning victories. At Lake Trasimene, on 21 June, 217 BCE, 25,000 Roman soldiers came to rout Hannibal and his 35,000 men out of the woods. Underestimating Hannibal, the Roman consul, Gaius Flaminius, who was heading the Roman forces there, ordered his forces to wait out the night before attacking the forces of Carthage.

During the night, Hannibal's men lit as many fires as they could near the lake, tricking Flaminius into believing that he was retreating. At dawn, Flaminius and his men marched into the woods, expecting to easily rout out a few stragglers left over from what they believed to be a rag-tag, amateur, defeated army. What they failed to know was that Hannibal's army had not retreated during the night and was, instead, poised just yards away from the Roman forces, ready to pounce.

As the Roman forces walked in a line along the beach, ready to march into the woods, Hannibal brought his cavalry around both sides of the Romans, blocking their escape. Then, his army quickly pounced upon the Romans, who were now fighting a three-front battle, with no chance of escape over the lake. The Romans there were doomed. Over 16,000 of Flaminius' men died there, most before they knew what was happening. The consul Flaminus was among those killed by Hannibal's army.

Fresh from his victories, Hannibal crossed into the Roman provinces of Picenum and Apulia and then onto the highly fertile area of Campania. Rome sent in a new general to take command of the forces opposing Hannibal. This general, who went by the tongue-twisting name of Quintus Fabius Maximus Verrucosus, left for Carthage in 217 BCE to demand an answer from their Senate as to whether or not they endorsed the actions of Hannibal. He stood in front of the Senate, grabbing two folds in his toga with his hands.

He asked them "I have two folds in my toga. Which shall I let drop – that holding peace or that with war?"[1] The Senate advised him to drop whichever fold he cared to and Fabius let go the fold of war. Despite this grandiose act, he adopted a cautious approach to dealing with Hannibal, by refusing to fight any decisive battle against the Carthaginian commander-in-chief. This allowed the Romans to rebuild their decimated forces while they harassed the logistics and supplies of Hannibal. With the threat from Hannibal, the Romans elected Fabius dictator in 217 BCE . Because of his prudent approach toward the Carthaginian general, the Romans also

[1] Philip Matyszak. "Chronicle of the Roman Republic." 2003.

gave him the nickname of "Cunctator" or "The Delayer."

In 216, Hannibal moved his forces to Cannae (currently Barletta, Italy), a town 400 km (250 miles) east of Rome. With his successes, Hannibal, now just thirty years old, had a growing number of allies and nearly 50,000 men under his command.

Here at Cannae, the long-awaited decisive battle finally took place. In what remains to this day the single greatest number of deaths during battle in a single day, Hannibal destroyed the assembled Roman army, killing 60,000 and taking 19,000 prisoners. Hannibal had lost nearly 10,000 men at this battle, but Rome was now severely weakened, demoralized and Hannibal stood just one knockout punch away from destroying Rome forever.

However, he needed reinforcements and he had no siege weaponry. He sent messengers to Carthage to request the materials he needed to destroy Rome and with it, ensure the power of Carthage throughout the Mediterranean. However, the Carthaginian Senate did what governmental bodies do best and delayed sending materials for years as they considered motion after motion, bills and amendments and this bureaucracy droned on while Hannibal continued to fight to a standstill on the Italian Peninsula.

An image of Hannibal Barca, the great Carthaginian general who nearly conquered the Roman Empire.

Public domain photo

In the campaign that would be the last serious attempt to destroy Rome from outside of her borders for over 300 years, Hannibal marched upon Naples, given the supplies he had available, but failed to take the city.

Hannibal had won control of the wealthy city of Capua after the battle of Cannae and this is where he chose to spend the winter of 216-215 BCE. The city was at one end of the Appian Way, constructed nearly a hundred years before; Rome was at the other end. At the time, Capua was considered one of the wealthiest and most influential cities on the Mediterranean and had been loyal to Rome since 338 BCE, but the demand of Capua for equal representation in the Roman Senate was denied and the Capuans welcomed Hannibal and his men. However, many of Hannibal's troops (likely non-Carthaginians, considering the affluence of Hannibal's home city) quickly became disenchanted with the wealth of the city's inhabitants.

Still waiting for reinforcements from Carthage, in 211 BCE, Hannibal marched upon Rome, attempting to deal the knockout punch. However, the fortified Roman positions held their ground and Hannibal's attack was repulsed.

Soon afterwards, the Romans began a long and desperate siege of Capua, to which the city finally capitulated. The 30,000 infantry and 4,000 cavalry the city had been able to lend to Hannibal were lost to the cause. The Romans quickly stripped civil liberties in the city, disassembled her magistrates, communal organizations and they city's ruling bodies. Then, the territory became property of the Roman Empire for selling and renting to the highest bidders and for various political favors, for years to come. The losses at Capua and Naples caused many of Hannibal's allies on the Italian Peninsula to abandon him and no longer would Hannibal be able to replenish his forces from their populations.

Finally, after four more years of fighting, with his forces depleting and Carthaginian power on the wane, Hannibal called upon his brother Hasdrubal, with whom he had fought on the Iberian Peninsula over twenty years before. Hasdrubal marched with his men, to meet Hannibal, but Hasdrubal was killed (reportedly by the Roman Consul Gaius Claudius Nero) in a battle upon the Metaurus River.

In 202 BCE, the inevitable occurred. Rome launched an invasion of Carthage, led by none other than Scipio Africanus the Elder, who had fought and lost to Hannibal at Trebia. Hannibal was recalled to Carthage to lead the defense of his home city. The showdown was to occur in Zama, North Africa, but when the armies met on the field of battle, Hannibal's raw, inexperienced troops fled, many defecting to the Romans. After that battle, or lack thereof, Carthage surrendered, and the Second Punic War ended with the complete submission of Carthage to the power of Rome.

However, even that was not enough to keep Hannibal from striving for the destruction of Rome. He quickly held what control he could in Carthage, rewrote the constitution of the city and re-organized the finances of Carthage. He also began making plans for a new, coming attack. Nevertheless, like Hitler in the closing days of World War Two, planning a new offensive with American and Soviet tanks already in Berlin, Hannibal's plan was too little, too late. The bureaucracy of the Carthaginian Senate had long before sealed the fates for both Hannibal and the formerly prosperous city. The Romans charged Hannibal with attempting to break the peace treaty, Hannibal was forced to flee from Carthage and he took shelter in Syria. One last time, Hannibal would fight the Romans, this time with Syria as his ally, under their king, Antiochus III. But after their twin armies were defeated at Magnesia in 190 BCE, Antiochus pledged to surrender Hannibal to the Romans.

Finally, Hannibal fled to northern Asia Minor, freelancing his military expertise to anyone who was fighting the Romans. Hannibal had one more defeat coming for the empire which he so detested. For a coming sea battle between the forces of Bithynia (located on the southern coast of the Black Sea) and Rome, Hannibal knew that the heat and close quarters aboard Roman ships would mean that the sailors would be huddled together nearly naked during the battle. Hannibal suggested the use of clay spheres, filled with live poisonous snakes, which would be lobbed onto the Roman warships. The trick worked and the Romans quickly fled the scene of the battle.

That was enough for the leadership of Rome and they decided to hunt him down at his villa. Past his prime, with his dreams and glory in shatters and the Roman army surrounding him, Hannibal poured poison into a flagon of wine and sat down for his last drink. His last wish is that the citizens of Rome could rest easy now, since they were so impatient for his death.

Thirty-five years after the death of Hannibal, the Romans besieged the city of Carthage for three years, using heavy siege weaponry including the *onager*, a heavy catapult that launched massive round stones at the city. The Romans finally took the once prosperous metropolis. Here, the soldiers of Rome slaughtered 450,000 men, women and children and sold the survivors into slavery. The city was burned, plowed over with salt and a curse was placed on the land directed at anyone who would ever try to rebuild Carthage.

By 170 BCE the Roman government, which once granted subjugated people and the population of client states full citizenship changed their policy. The citizens of small allies on the Italian peninsula no had no representation within the Roman government. This was a mistake that would cost Rome greatly in the near future.

Change was also occurring in the Roman legal system at the same time. The office of tribune was expected to be filled by a person who was close to the people, leaving his doors open for travelers who needed a meal or a place to sleep. In 124 BCE, Gaius Gracchus was elected to this post, to serve for the following year. He was a thoughtful, calculating person, yet he was quick to anger. Gracchus had a slave listen to him as he engaged in politics so that the slave might play a flute whenever his anger would start to rise to the surface. He re-arranged the tax system, giving out contracts for tax collection to large businesses that would collect the taxes, taking a share of the money they collected. Most other functions of the government of the republic including administering the aqueduct system, roads and supplies were supplied by these people and organizations known as *publicani*. Gracchus forbid Senators from participating in the business of tax collection, or from sitting on juries in cases of extortion. The Roman knights controlled tax collection, and with Senators forbidden from involving themselves in extortion trials, a massive shift of power away from the Senate and to the knights occurred. The Roman Senate had now lost much of the power that they had held throughout the history of Rome.

A paradigm shift in how armies dealt with the spoils of war also came during this time. The Greek armies in the past were under orders that they "[C]ould dispose of the proceeds from the sale of booty in various ways... but whatever was brought back became the property of the state."[1] Similarly, the Romans followed the same principles *until the war against Carthage.* Likely enraged by the amount of wealth

---

[1] Finley, M.I. "The Ancient Economy." 1973

being accumulated by legislators who took land through confiscation (especially in Italy), as well as by the great opulence of Carthage, the soldiers and commanders began to keep the "booty" for themselves. This was a new way of doing things. Cities had always been plundered during war since the beginning of time. But this was far more; for the soldiers and commanders were no longer pillaging and foraging for some leader in a distant city, selecting a few prize gifts to send back to their capital. They were working for themselves now and the conquered lands would pay the price.

Back in Alexandria, Eratosthenes was doing work in prime numbers, attempting to find a prime number theory; a formula that produces only prime numbers. He did not succeed in that quest; in fact, even today's mathematicians have not found such a formula. However, he developed a process known as the Sieve of Eratosthenes. The process is described in the box below:

The Sieve of Eratosthenes

1 **2** **3** ~~4~~ **5** ~~6~~ **7** ~~8~~ ~~9~~ ~~10~~
11 ~~12~~ **13** ~~14~~ ~~15~~ ~~16~~ **17** ~~18~~ **19** ~~20~~

This process finds prime numbers from one to any number one wishes.
Simply write down the numbers (in this case, 1 through 20) and we know the number one is a prime (shown here in bold letters).
Go on to the number two. That is also a prime number. Now cross out every number divisible by two (represented here by a double strikeout).
Then, the next number, 3, is also a prime. Cross out all numbers divisible by 3, (here 9 and 15, with a single strikeout).
Four is already out, since it was divisible by 2.
Five is the next number and the next prime.
Continue this way until you have reached 1/2 of your highest number and the only numbers left will be prime numbers.

In 205 BCE, less than ten years before the death of Eratosthenes, Ptolemy V Epiphanes (the Illustrious) came to power in Egypt. He quickly had his foreign possessions divided between Macedonia and the Seleucid Empire, which greatly hurt the power and prestige of Egypt.

One of the most intriguing books on mathematics in the ancient world was written about this time – "On Conics," authored by Apollonius of Perga. He was also working at the great library and wrote this work on conic sections. In fact, it was he who first developed the names ellipse, parabola and hyperbola for the three distinct types of cuts which can be made through a cone.*

Apollonius was considered such a mathematical genius that the people around him gave him the moniker of "The Great Geometer". In addition to his work with conic sections, he also studied irrational numbers, optics and mathematical astronomy.

Apollonius was born in Perga (current day Murtana, Turkey) about the year 262 BCE. The people there worshiped the nature Goddess, Artemis, and the city was a center of culture. In his youth, Apollonius traveled to Alexandria to learn under the mathematicians at the Great Library. Soon, he would begin teaching there as well. During his career at the library, he visited the competing library at Pergamum, likely

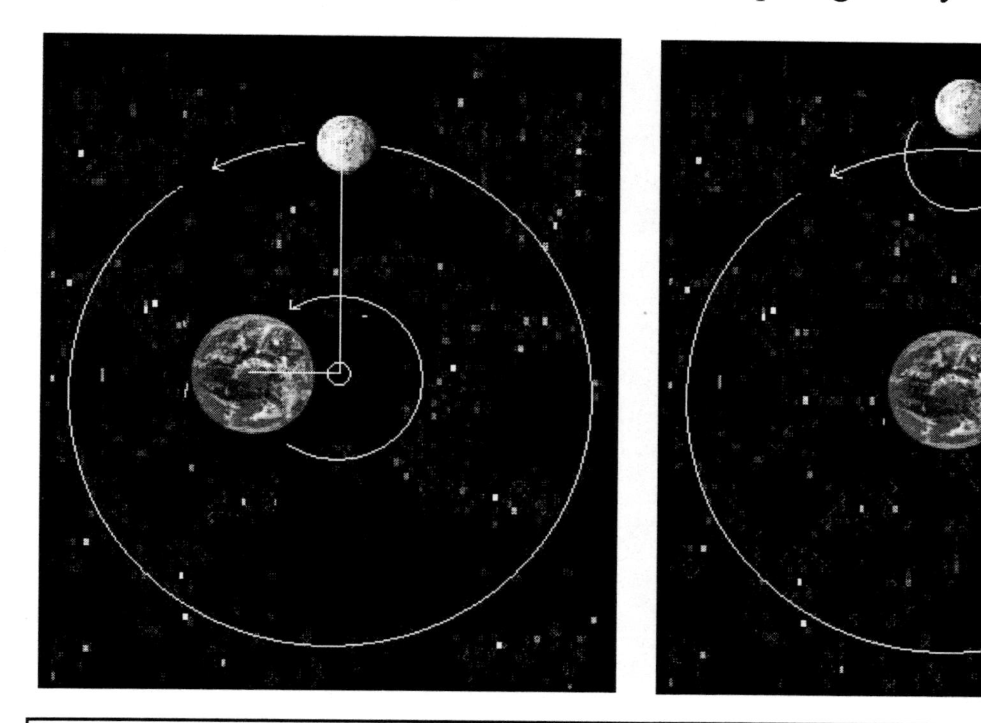

Epicycles and eccentric orbits:
These two ideas were developed to explain the apparent backwards motion of the planets called retrograde motion. With an eccentric orbit (left), both the Earth and the other planet revolve around an invisible point at the center. In the system of epicycles (right), the other planet revolves around both the Earth, as well as its own mini-orbit, known as an epicycle.

to exchange ideas and theories with the scholars of the land. He was the first person to calculate that rays from the Sun could not be focused to a single point by a spherical mirror and determined that a parabolic mirror would be required to accomplish this feat. Although others had earlier attempted to explain the apparent retrograde motion of planets through the use of epicycles and eccentric motion, Apollonius was the first person we know of to have made mathematical descriptions of these systems in great

* A circle can also be created in such a manner, but circles are a type of ellipse. For a diagram of conic sections, see the chapter on Hypatia.

detail. He wrote a total of eight books before he died in Alexandria in 190 BCE.

Another person who deserves credit and is lost to history was named Ctesibius, who worked with pneumatics in Alexandria and invented the fire engine (likely slave-powered) along with greatly improving water clocks. Ctesibius was two years younger than Archimedes and was born in a suburb of Alexandria, the son of a barber. He worked on devices that ran on compressed air, including siphons, pistons and developing a set of adjustable powered mirrors for his father's barbershop.

In another gift to our age from this era, Ctesibius also invented a new musical instrument. This instrument became known as a *hydraula*. It used water to drive air through pipes and the pipes were controlled by means of a keyboard and pedals. In short, the father of hydraulics also invented the pipe organ.

Rome was now a rapidly growing city where the poor tended to live in large apartment buildings, called *insulae*, of three to eight stories each. Homes for the wealthy would often open into an enclosed patio area where food was prepared, upon a large black stone. The Latin word for black was *ater*, so these rooms provided the basis of our modern word "atrium." Many of the noble families who lived in these large homes would rent out the street frontage of their homes to small businesses, whose owners lived upstairs. Most people also worked from home in other rooms which surrounded the atrium. Working from home is not a new concept! Slaves were often expected to share small rooms built under the floor of the main house.

The new musical instrument that appeared at the end of the third century BCE, the hydraula, or pipe organ.

Public domain photo

About this same time, the first commercial bakeries appeared in Rome. Most families still chose to bake their own bread, but they could also now buy it from a commercial bakery for the first time. Other restaurants, serving inexpensive food quickly to the working poor living in the numerous apartment dwellings, also sprung into existence. This legacy for the modern age from life in second century BCE Rome is what we now call fast food.

Frail and blind in his old age, Eratosthenes ended his days in 196 BCE, starving himself to death. Three years later, Ptolemy V married the Seleucid princess, Cleopatra I*.

* Cleopatra I was a princess of the Greek Seleucid dynasty, founded by Seleucus I, which ruled over

Hipparchus of Rhodes 190 BCE – 120 BCE

Hipparchus of Rhodes

Photo believed to be public domain

Born in the city of Nicaea, in Bithynia (current day Iznik, Turkey), Hipparchus would become a great astronomer and mathematician. He did much of his observations at Alexandria and continued his work when moved to Rhodes towards the end of his life.

In astronomy, Hipparchus created a star table containing over 850 stars due to a rare event in the heavens which occurred during his life. A "new star" was observed in the constellation of Scorpio, something that shouldn't happen if the heavens were constant and unchanging. Hipparchus could not be sure if the new star was truly a new star, or if he had just overlooked the star in his previous observations. Therefore, he decided to catalog the stars, in hopes that should this happen again, he could be certain that he had not previously observed the star in question. We now know that what he had observed was a supernova – a super massive star dying in a final flash of glory, brighter than a hundred billion suns. What Hipparchus had believed to be the beginning of a new star, was in fact, the end of the life-cycle of a distant star that had previously been too dim to be observed.

He also discovered the precession of the equinoxes. Precession is simply the slow moving of the Earth's axis in a circular motion around itself. This is similar to a child's top spinning and slowing down as forces other than the original spin begin to become more prominent. The top creates a "cone-like" rotation around itself as the energy of the spin begins to die out. The Earth's precession works in a similar fashion, rotating once every 26,000 years or so. It is due to this precession that from 2700 to 1400 BCE, Alpha Draconis, in Draco, was the North Star and Polaris will no longer be the North Star in only a few thousand years. His calculation of forty-six seconds of arc per year (each second of arc being 1/3600th of a degree) was much better than the 36" (seconds of arc) calculated by Ptolemy almost three hundred years later and surprisingly close to the modern accepted figure of 50.26".

Hipparchus is often credited with the discovery/invention of trigonometry and the dividing of a circle into 360 degrees. Each time we talk about an angle as being "so many degrees" we are using the system devised by Hipparchus. To further aid the accuracy of his astronomical observations, Hipparchus improved the main

much of the Aegean world from 312 to 64 BCE. The more famous (and last) Cleopatra, famous for love affairs with Julius Caesar and Marc Antony, was Cleopatra the VII (69 – 30 BCE), who ruled Egypt (off and on) for twenty-one years.

astronomical instrument of the time, the *Dioptra*. This was very much like the equatorial mounts found on many modern telescopes, with a simple tube and pointer in the place of the telescope.

Ptolemy V, king of Egypt, died in 181 BCE, leaving the country in the hands of his five-year-old heir, Ptolemy VI Philometor ("Loves his mother").

Four years later, the Romans, under their new consul Tiberius Sempronius Gracchus, took control of Sardinia, which lies south of Corsica. The number of captives taken as slaves on this island was so great that Romans began to use the phrase "Sardinians for sale" in the sense that we today would say "A dime a dozen." A tradition followed by many people in Sardinia was to poison elderly relatives who the family had considered had outlived their usefulness by feeding them belladonna. This toxic family of plants (except for tomatoes) kills people, leaving their face contorted into a twisted smirk. This is the origin of our modern phrase "sardonic grin."

Less than ten years later, as Hipparchus reached his twenties, in 168 BCE, Egypt was invaded by another Macedonian, Antiochus IV, who took Ptolemy VI prisoner in the process. Antiochus gave the throne of Egypt to the younger brother of Ptolemy VI, who became Ptolemy VII. Rome sent an ambassador, Popillius Laenas, to order Antiochus to withdraw his forces from Egypt. The two met on a beach and Antiochus told the Laenas that he needed time to consider the proposal. Laenas drew a circle in the sand around the Macedonian king, telling Antiochus that he could leave the circle once he had decided. This event was the original incident that we refer to every time we use the phrase "A line in the sand." When the forces of Antiochus withdrew soon thereafter, the two brothers reigned for a short while as joint kings, along with their mother, Cleopatra the Second, but discord quickly grew in the empire.

After arbitration between the Ptolemys from Rome in 163 BCE (at the request of Ptolemy VI the previous year), Ptolemy VII was given Cyrenaica, a fertile area in what is now NE Libya as his land to rule. The Seleucid king, Demetrius II, died in 150 BCE and the kingdom of the Dynasty was offered to Ptolemy VI, who turned down the offer of the throne. Five years later, he died in battle defeating the greatest foe of Demetrius II, Alexander Balas, who also died at the battle. Alexandria and all of the lands of Egypt were left in the hands of Ptolemy VII.

Hipparchus was forty when, in 149 BCE, the last of the Punic Wars broke out. Marcus Porcius Cato, also known as Cato the Elder, a Roman statesman with the desire to destroy Greece, went to mediate talks between Carthage and Numidian tribesmen. The wealth and affluence of Carthage enraged Cato and he began to end every speech with the words "*Delenda est Carthago*" (which translates as "Carthage must be destroyed."). Soon afterwards, a member of the Scipio family, who opposed a war with Carthage, began to end his speeches with "Leave Carthage alone."

A minor treaty infraction gave the Romans the pretext they needed to attack Carthage and within three years, after fierce house-to-house fighting, Carthage was destroyed and the remaining inhabitants were sold into slavery. The remains of the city were burned and then the fields were plowed with salt and a curse was placed upon any person who would dare to try to ever rebuild the city.

The Romans had now begun to engage in serious trade with other lands around the Mediterranean. Most oil lamps in the area were now produced by just a few businesses in Rome and the surrounding territories, financed largely by wealthy Roman senators. Goods such as oils and wine were shipped in clay jars known as *amphorae*, many of these goods coming in and out of Rome went through a port city on the Tiber River called Ostia. By 100 BCE, Rome had passed a law that forbid senators from engaging in trade or even owning ships that could be used for large-scale trading. These laws were easily circumvented through the use of middlemen known as negotiators, who would engage in the trade themselves, using money supplied by senators.

In 146 BCE, the Roman Empire conquered Corinth after which Greece and all her lands came under the power of Rome. However, the Romans as a whole always had a special admiration for Greek culture and the new colonies became influential in religion, literature and philosophy. Not everyone was as pleased with the teachings of the Greek philosophers as others, and some people resisted the new philosophy that they knew would pervade Rome from Greece. Even before the takeover, as the influence of Greece was beginning to be felt in Rome, Cato the Elder wrote:

"*I shall speak about those Greeks in their proper place... They are a depraved and unruly people, and I prophesy that when that nation gives us its literature, it will corrupt everything. All the more so if it sends doctors here. They have conspired to kill all barbarians with their medicine, and they even charge us for doing so! They regularly call us barbarians and hurl more filth at us...*"[1]

Cato led a group of people who fought Grecian influence in the Roman Republic, even refusing to let his sons visit doctors, since they usually came from Greece. Toward the end of his life, however, even Cato began to learn the Greek language.

The next year after Rome conquered Greece, Ptolemy VII became Ptolemy Euergetes (the "Benefactor") II in Egypt. After the death of his brother, with whom he had co-ruled Alexandria, he married his brother's widow, a woman named Cleopatra II. This marriage quickly went rocky and the couple split ways. After this divorce, Ptolemy VII married his ex-wife's daughter, also named Cleopatra (Cleopatra III). This marriage would eventually cost Euergetes three years of rule in Egypt.

Eleven years later after Rome won control of Greece, the Roman Empire became engaged in what would be the first of many civil wars, the First Servile War, which began with a slave uprising on the island of Sicily.

---

[1] Pliny the Elder. "Natural History." 29.7.14. Quoted in Philip Matyszak. "Chronicle of the Roman Republic." 2003.

After the fall of Carthage, Rome took the lands on Sicily that had belonged to wealthy Carthaginian landowners and made them available at a low price to Roman citizens. This caused a huge rush of Romans to Sicily, where they set the slaves to work on producing tremendous amounts of grain to be sent to Rome. Many of these new masters did not feed their slaves well and the over-crowded, malnourished slaves began to indulge in robbery from poor Sicilian families in order to provide for themselves and their families. Content with the idea that their slaves should live at no expense to their owners made the slaveholders turn a blind eye to the corruption and theft occurring throughout the island.

Many of the slaves became enamored with one of their own members, a self-described prophet and conjurer from Syria named Eunus. He collected 120,000 followers amongst the slaves on the island and he led a slave rebellion in 135 BCE. This would become known as the First Servile War and two more would follow over the next sixty years. Eunus, together with a Cilician name Cleon, fought the Romans successfully for three years, until 132 BCE, when Cleon was killed in battle and Eunus captured. Eunus died in captivity before he could be brought to be tried before the Roman courts. Rome had crushed her first major slave rebellion.

Hell hath no fury like a woman scorned, particularly ones named Cleopatra and bitter of the marriage, the elder Cleopatra led an army against Ptolemy VII and drove him from Alexandria in 130 BCE. Three years later, Ptolemy VII returned to Alexandria and, according to Greek records of the time, was a harsh, despotic ruler, although he is credited with administrative reforms in government and he was a large benefactor to religious institutions of the land. However, he also drove many scholars from the Great Library.

His son with the younger Cleopatra, Ptolemy VIII (Ptolemy Lathyrus, aka Soter II) ruled in conjunction with his mother, who held the real reigns of power. She eventually forced him to accept his younger brother, Ptolemy Alexander (who would grow to become Ptolemy IX) as a co-leader in later years.

Upon his death, Ptolemy VIII left the fertile area of Libya to his illegitimate son, who became Ptolemy Apion (died in 96 BCE). Egypt and Cyprus were left to the second of his wives, Cleopatra III, who was directed to select one of her sons as a co-ruler of the land. The Ptolemaic Dynasty was nearly ended; what had begun with the height and glory of Alexander and Ptolemy I, began to fall apart into chaos and struggles for power.

Of Love and War 120 BCE – 48 BCE

For the next one hundred years, science remained nearly stagnant. Effort was put into expanding and preserving the work done earlier in the lands surrounding the Ionian Sea, as well as in Alexandria. The first century BCE also saw what were perhaps the single greatest series of love affairs and some of the most famous battles in history. It was also the time of what was perhaps the greatest invention of the ancient world.

Cleopatra VII, from "Cleopatra and the Peasant" by Eugene Delacroix, 1838.

Public domain photo

Sicily once again erupted in a slave revolt in 104 BCE, this one far more serious and widespread than the revolt over thirty years earlier led by Eunus and Cleon. This would become known as the Second Servile War and once again, Rome was victorious.

Thirteen years later, in 91 BCE, the formerly faithful Roman allies began to fight against Rome, at least partly to regain Roman citizenship for all people of the Italian Peninsula.

The so-called "Social War" of Rome began in 89 BCE, when dissent within the Italian Peninsula began to become ever more fervent, as the cities outside of Rome demanded a say within the Roman government. Hostility led to rebellion and Rome found herself surrounded by angry subjects, demanding representation. The army was divided, as the ranks of the army were made up largely of the *capite censi*, who were more devoted to their individual commander than to Rome itself. Rome would put down these riots, after a bitter, hard-fought war on all sides and after giving the rebels what they really wanted – Roman citizenship.

Three Roman generals would become known for their dedicated and successful service to Rome during these years: Gaius Marius, Gnaeus Pompeius Strabo* and Lucius Cornelius Sulla. The fame and power that Sulla received from fighting in the Social War would propel him to the office of consul after the end of that war. He had golden hair and striking blue eyes, but also had a terrible complexion that broke out into red blotches when exposed to sunlight. Sulla had not always been wealthy, having lived during part of his youth in an apartment block downstairs from a freed slave. Sulla loved the arts and kept company with dancers and actors, eating fine food and had a great personal charm which certainly endeared him to the people who elected him to the office of consul in 88 BCE. While in Asia minor in ten years earlier, Sulla met a soothsayer who forecast to Sulla that he would rise to become one of the most

* Strabo was the father of Pompey the Great.

powerful men in the world and die at the height of his success.

After the election to consul, Sulla was beginning the height of his power, at the command of a vast army, poised to go back to Asia Minor (where had been stationed before the Social War) to defeat King Mithridates, reclaiming Roman land taken by Mithridates there and in Greece.

The effect of non-Roman people from the Italian peninsula serving in the Roman legions for the first time also changed how the army behaved. No longer a group of Roman citizens who would march upon orders from Rome without question or dissent, units were now ready to march into war on behalf of individuals within the government.

Mithridates VI of Pontus, one of the greatest enemies of Rome.

Public domain photo

The power and reach of Rome was becoming ever more prevalent and the ancient country of Pontus, on the Black Sea, was the last to seriously challenge Rome for power in the world. The leader of Pontus, Mithridates VI, had extended control of the kingdom beyond the Caucasus and Rome began to feel threatened. Mithridates had spent most of his life before he achieved the throne as a fugitive. He killed all of his brothers and then married his sister so that he could be assured of becoming king.

In 88 BCE, the two lands went to war in the First Mithridatic War and Pontus quickly conquered nearly all of Asia Minor that first year. The following year, a power-hungry young Roman lieutenant by the name of Lucius Murena attacked Pontus without justification and within two years was defeated by Mithridates. Aulus Gabinius replaced Murena in his command under Sulla and negotiated a peace with Mithridates. In 85 BCE, the Roman general Fimbria marched his troops into Asia Minor and defeated Mithridates, who also saw his armies destroyed in Greece at nearly the same time. In the treaty to end the war in 84 BCE, Mithridates paid a hefty fine to Rome and surrendered much of his kingdom.

The land of Bithynia (now part of Turkey) had been given to Rome as an estate by the king of Bithynia, Nicomedes III (who had been a close ally of Rome), when he passed away in 74 BCE. Mithridates was determined to strike at the increasing power of Rome and did so that year, marching his troops through Asia Minor.

Sulla began to prepare an army to invade Bithynia and return the area to Roman control. However, after Sulla left Rome, the now elderly Marius had a tribune named Sulpicius Rufus sponsor a bill in the legislative body to award the command of these forces to Marius. Going well beyond political infighting, the allies of Marius began rioting in the streets to the point where Sulla himself took refuge in Marius' home, likely reasoning that the mob looking for him would never look for him there. Marius took command of Sulla's forces for a short while, but the troops remained loyal to their former commander, stoning to death the tribunes sent to them by Marius.

Sulla soon appeared to them, speaking of treason and the mob rule of Marius running affairs in Rome. He knew his only chance to remain in power was to march upon Rome with his forces that had been assembled for the planned invasion of Bithynia. He tried to rouse his troops, but many of them were fearful of beginning what they knew would be the beginning of a Roman civil war. Although all but one of his officers left his side, refusing to fight, Sulla was determined to take the city by force. The next pair of emissaries sent by Marius to negotiate with Sulla returned to Rome stripped of their robes and with their staffs of office broken in half. Sulla finally took Rome itself and elections were held for the following year.

An image of Lucius Cornelius Sulla Felix, taken from a Roman coin.

Public domain photo

Elected as praetor, Cinna swore an oath to Sulla not to attack him or persecute him for what turned out to be the exceptionally unpopular act of taking Rome by force. Cinna soon reneged on his promise and ordered that Sulla would stand trial for this crime. Ignoring the summons, Sulla marched his forces into Greece to successfully fight against Mithridates.

Meanwhile, ready to march back to Bithynia, Rome countered Mithridates with an army headed by Lucullus, a favorite general of Sulla's, assisted by Caius Aurelius Cotta. Mithridates began to defeat Cotta, but found himself surrounded from behind by Lucullus, who demolished the Pontic army. Mithridates fled back to Pontus, where he stayed for one year until he was forced to flee to Armenia, where he lived until the Romans attacked there in 68 BCE. Returning to Pontus, Mithridates took refuge in his capital city until Pompey, who had replaced Lucullus in 66BCE, forced Mithridates into exile in the Crimea, where he ordered a slave to kill him three years later.*

After the battle against Mithridates, Sulla began to march his army onto Rome. The Roman allies were fearful that their citizenship, which had been won during the Social War, would be lost under Sulla. To ease their opposition, Sulla played both sides of the dispute, promising the people of the peninsula that he would not wage war against them, while he savagely destroyed those who opposed him. The last battle before he took Rome was just outside the city, at the Colline Gate, where Crassus won the battle just as Sulla was beginning to plan for defeat. Taking Rome, the Senate

---

* There is a legend that Mithridates regularly ingested small amounts of poison in order to build up an immunity against the substances. Supposedly, he attempted to commit suicide by poison and was unable to ingest enough to harm him, so he chose to die by the sword instead.

declared him dictator – the first one Rome had seen for 120 years. Sulla immediately began a purge of the Senate, killing anyone who opposed him and even, at one point, posting a list of those to be executed. Political and financial gain also increased the number of people on the death list, as senators who owned properties desired by the allies of Sulla were added to the list.

However, far from the tyranny that was expected by many in Rome, Sulla worked to expand the representation of the people in their government, adding more magistrates, quaestors and praetors. In 80 BCE Sulla voluntarily ended his own dictatorship, fearing what would happen to the Roman people if he did not step down. The Republic had been restored, but there was now a new reality in Rome. Sulla had shown that any leader with a military force behind him that was large enough could overthrow the Republic if he desired. Sulla had chosen not to do so, but people in power were now aware that such a feat could be accomplished, and that is exactly what would happen.

Now at the very height of his power and fame, Sulla was now happy as well. The prophesy given to him years before that he would die at the height of his power and happiness began to weigh on his mind. His beloved wife, Metella, died of the plague soon after delivering a set of twins. His happiness, so recently obtained, was now shattered. He set about compiling his memoirs and died two days after they were completed, thereby fulfilling the prophesy. Two years later, in 78 BCE, Pompey killed the father of the infamous Brutus when the two engaged in battle in Gaul.

A few years before the death of Nicomedes III and the ceding of Bithynia to Rome (in about 80 BCE), an unknown scientist/philosopher created what was, more than likely, the greatest invention in the ancient world.

Nearly two thousand years later, in the year 1900, a Greek sponge diver by the name of Elias Stadiatos was diving just 30 meters (about 100 feet) off the coast of the tiny island of Antikythera and 60 meters (two hundred feet) beneath the water, when he discovered an ancient shipwreck. Aboard the wreck, Stadiatos discovered statues*, pottery, jewelry, wine, furniture and fine bronze pieces. After further exploration by this group, along with archeologists and the Greek navy, they also found far more of these treasures, along with a couple of overgrown green lumps of metal and the sea life which had made it their home. It was the little green lumps of metal that would prove most interesting find of them all. After examination, it appears, from the pieces that remained from the device, that this was an analogue computer from the first century BCE.

The part of the device found appears to have been able to calculate the position of the Moon and Sun, along with the positions of Mercury and Venus. It could do this with any date entered into it by the operator, using interlocking differential gears. The

* These were Greek statues, from the fifth and sixth centuries BCE, which had been stolen from the Greeks by the Romans in the first century BCE and may have been stolen back just before the sinking of the ship.

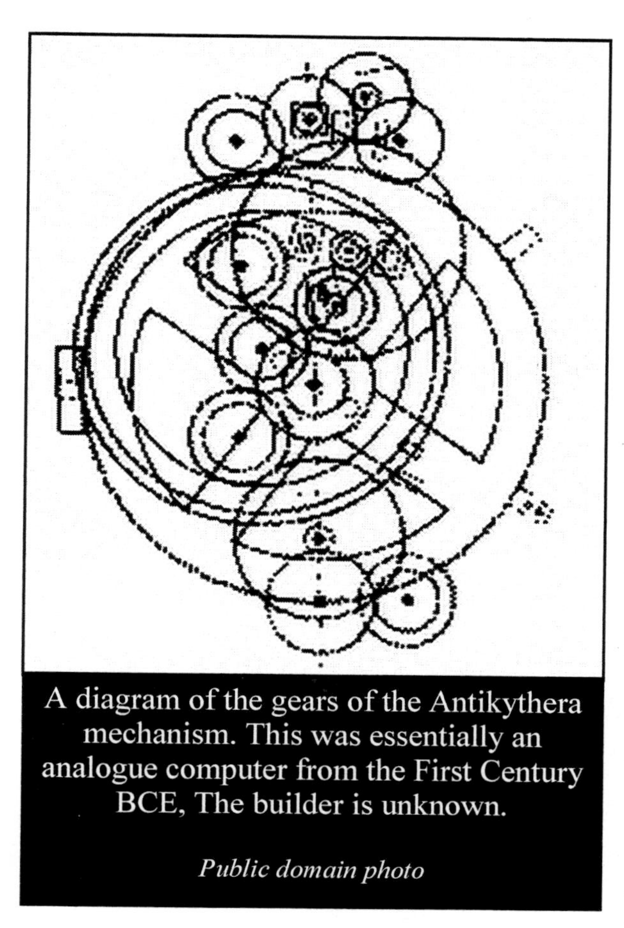

A diagram of the gears of the Antikythera mechanism. This was essentially an analogue computer from the First Century BCE, The builder is unknown.

Public domain photo

way the gears are constructed, it would also be possible to attach an additional similar box to the part that was found, in order to calculate the positions in the sky of Mars, Jupiter and Saturn. The maker or makers that created this anachronous device even added a slip-gear within the workings to account for leap years.

The closest modern equivalent to the level of technology that the Antikythera represents would be the Norden bombsight, which was the second most guarded secret of the Second World War, behind only the atomic bomb.

The Antikythera used thirty-two gears and had doors and levers on each side, along with bronze pointers above a circular dial. These were marked with the signs of the zodiac, like the numbers around a clock face.

What is more amazing than the Antikythera itself are hints in the historical record that this device was not the only one of its kind in the world. Perhaps there still exists somewhere, underground or underwater, an even more advanced 2,000-year-old computer waiting to be discovered. The ancient world may have been witness several of these machines. Nobody knows who built the device or devices, though it is more than likely it was developed in Alexandria, perhaps by one of the scientists working there at that time whose works are now lost to the world.

The third of the three Servile Wars struck Rome in 73 BCE and would become the best known of the three Servile Wars. For this rebellion, which began with an escaped gladiator who gathered an army around him and fought the Romans for two years, was led by a man named Spartacus.

Spartacus had been a gladiator from Thrace when he escaped from Capua[*] in 73 BCE and he quickly fled to Mt. Vesuvius where an army grew around him. Seventy thousand men, women and children would find their way under the protection and leadership of the escaped gladiator.

He began to bring his group northward, in an attempt to escape the Italian Peninsula and with it, the authority and power of Rome. Time after time, the Roman

---

[*] Capua was a city in southern Italy, north of Naples. It was strategically important due to being located right on the Appian Way, the main road to Rome.

legions tried to stop this group from their migration towards freedom and each time, the military tactics and planning of Spartacus stopped the Romans in their tracks. His forces held the upper hand throughout Southern Italy and they began to gather wealth and riches from the areas and people over whom they ran.

This turned out to be the lynchpin in what would be the failed escape of the group from the Italian mainland. For once the people under Spartacus began to get a taste of riches, they wanted even more wealth. Thus, rather than continue north to safety, they again turned southwards to gather more plunder.

Spartacus is finally defeated at the Battle of Lucania by Crassus. Pompey would finish off this slave rebellion.

Photo believed to be public domain

At first successful, Spartacus was about to meet his match. The Romans sent out a man named Marcus Licinius Crassus to defeat the one-time gladiator-turned-general. Crassus set up a trap on the "toe" of the Italian mainland, and Spartacus reached a deal with Cilician pirates for them to transport his troops to Sicily, where he hoped to organize and lead another slave rebellion against Rome. However, the pirates reneged on their promise to him, and they left, leaving Spartacus' army stranded on the Italian peninsula. Crassus then had his men build a wall across the peninsula, hoping to trap the rag-tag army and destroy them, but Spartacus escaped. Spartacus headed northward and was brought unwillingly to battle at the behest of his generals. Heading directly towards Crassus at the Battle of Lucania, Spartacus fell in battle, although his body was never found. The survivors tried to begin the trek northwards once again, but Pompey returned from his work in Spain and, along with Crassus, destroyed the last remnants of Spartacus' rag-tag army.

Many corpses and many of the captured members of Spartacus' army were crucified along the Appian Way. The order to remove the bodies never came from Crassus and in all, 6,000 bodies hung along the side of Rome's main road for years, as their bodies decayed. This was a not-so-subtle warning that any other slaves who attempted to revolt would be treated without compassion or mercy and quickly and painfully annihilated. The treatment of the Romans who had been captured by Spartacus was far different. The Romans rescued 3,000 of their men who had been captured by the former slaves and these men were unharmed.

The opening decades of the first century BCE also produced the philosopher Zeno of Sidon, who came from the area that is currently the Mediterranean coast of

Lebanon. He was an Epicurean, believing in pleasure and avoidance of pain and desire as the route towards achieving the greatest good in life. Zeno was a student of Apollodorus and taught in the same garden where Epicurus had founded his school of philosophy in 306 BCE.

Although Epicurus belittled science and mathematics solely for the sake of knowledge, Zeno questioned the work of earlier philosopher/scientists (particularly Euclid) with a more inquisitive, methodical method. One day, when he was speaking in "The Garden" he had a special guest in the audience – the historian and politician Cicero.

Tullius Cicero was from a wealthy, yet not aristocratic, family and showed a great deal of political ambition from an early age. As a student, he learned so quickly and was such an excellent student that he attracted attention from around the Republic.

It was quite common for Roman men to have both a given name (usually the same as their father's, no matter how many boys the family had* and to also be given a *cognomen*, a nickname, often slightly derogatory. Cicero meant "Chick Pea," Caesar was a name meaning "Curly" and Brutus meant (essentially) "Brute."

Given the great deal of power which was wielded by only a small group of people who held the real power during the Roman Republic, the only roads open to political office for Cicero were either the military (which Cicero despised), or through practicing in law. He chose the latter.

Cicero was elected on his first attempt for each of the four primary offices in Rome, including Consul, at the earliest age that he was allowed to run for each of the positions. Doling out political favors, Cicero had a unique tool in the law that allowed him to yield great power. Lawyers had not been allowed to accept fees legally in Rome (although this did happen from time to time) since 204 BCE. Under the reign of Cicero, the lawyers knew that throwing the decision of the courts to the whims of this powerful man would yield bountiful returns in the future.

A bust of Cicero

Public domain photo

In 63 BCE, while Cicero ruled as Consul, a collusion known as the conspiracy of Catiline was brewing, centered around of a financially ruined patrician (nobleman) by the name of Sergius Catiline. At first, Catiline tried to work within the Republic, fighting for relief of debts for the populous (and, incidentally, for his own family, the Sergians, who were in a period of a great decline in their wealth). They were supported at this time by none other than Julius Caesar. The Senate, who were

* This was a society full of "Juniors." Female children were often given the feminine version of the father's name, just to keep things simple.

mainly creditors, were opposed to the idea.

Pompey the Great

Public domain photo

This movement evolved into a plot to take over the Roman Empire and it was uncovered by Cicero's government. Five of the people involved in the conspiracy were ordered put to death without trial by Cicero, even after a call for leniency on behalf of Caesar. This act endeared Cicero to his people in the short-term, as he claimed that he had single-handedly saved the Republic from falling. However, this act would also come back to haunt him. It also furthered the dislike of Caesar by many of the Senators.

The Roman Republic (such as it was) would further continue to disintegrate into chaos in 60 BCE, when three men conspired to take control of the Empire and ruled as Rome's first Triumvirate. These men were two of the people who had helped to put down the uprising of Spartacus in the Third Servile War: Pompeius Magnus (also known as Pompey the Great), Marcus Licinius Crassus and a military leader by the name of Julius Caesar.

This event, perhaps more than any other in history, shows what can happen when people with tremendous wealth begin to hold combined power within the structure of government. For although Caesar and Pompey were military and political figures, they also needed the support of the wealthiest people of Rome to maintain that power.

Crassus, who owned half of Rome

Photo believed to be public domain

Crassus was the richest man in the Roman Empire, owning over half of the property in Rome and this secured for him a place in the first government after the fall of the Roman Republic. Crassus had begun acquiring great wealth when he was only a youth through the slave trade, buying up properties damaged by a series of fires which ravished Rome at the time* and through the mining of silver.

However, greed and a quest for ever-increasing wealth did not rest solely with Crassus. Back in Rome, another up-and comer named Brutus was lending tremendous amounts of money to governments, sometimes at extremely high interest rates, which was usually prohibited in the Roman Empire. One loan had been made to the city of Salamis, in Cyprus, by Brutus at a 48% interest rate five years before the death of Crassus. As expensive as the paybacks could be, governments under

* Many people in Rome at the time had accused Crassus himself of setting the fires in order to buy damaged properties at a reduced price.

Rome needed this tremendous influx of cash to feed an ever-growing bureaucracy. The Governor of Cilicia at the time, Cicero, was paid 2.2 million sesterces, which was about three and a half times what he himself stated could permit a life of luxury. Unlike the Greeks, the Romans did try, from time to time, to control interest rates, but the lack of any real sense of modern economic theories created a situation where the Roman government would know it wanted to do something about the economy, but was unsure as to what to do. Military expansion was usually seen as the quickest, easiest way to grow the economy, and most of Rome's leaders were former military personnel. It was nearly impossible to succeed in Roman politics without first establishing a successful military career. Social status could rarely be achieved without a record of military prowess and the economy of the Roman Republic became dependent, to a large degree, on continual expansion through the use of military force.

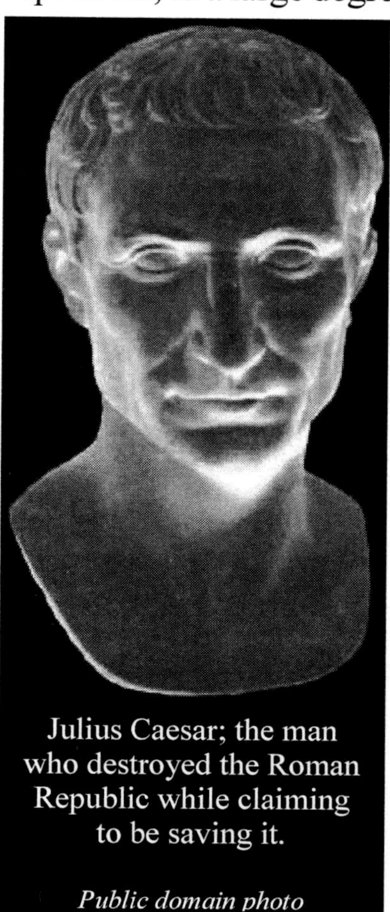

Julius Caesar; the man who destroyed the Roman Republic while claiming to be saving it.

Public domain photo

Political bribery, favoritism towards friends who helped elect people to office and collusion for power became the norm throughout the Roman political system. The system had become corrupt, lust for power among many candidates required money to achieve and that money was repaid with political favors, land and contracts. Many within the Republic were certain that the government was now run by a combination of wealth and military power, subverting the Republic. In many ways they were correct, but the cure would turn out to be worse than the disease.

Julius Caesar was from a family which actually pre-dated the city of Rome itself, the Julian family. His ancestors had once been greatly revered within the Republic, but had fallen into disfavor centuries before. However, the Julians had experienced a sort of revival in the decades leading up to Caesar's birth and his connection to the ancient Julian family gave him the impetus and credentials he needed to establish his own power within the Republic.

Caesar himself was a sober, calculating man who rarely gambled or drank alcohol, but he did have a passion for womanizing and that would bring him, later, into a relationship with the most powerful woman in the world. Caesar was also likely not as handsome as the busts and portraits he commissioned of himself, for he was also a ruthless self-promoter. He dressed well, in the style of an aristocrat, despite his family's lack of money. He had dark, intense eyes and wore fringes on his tunics and togas to further impress people with his style. However, he was also deeply troubled, suffering from massive headaches and terrible recurrent nightmares, perhaps due to his epileptic seizures, the cause of which was unknown at that time and that may have produced great fear of mortality within him. He also, apparently, had little sense of taste. Once, while at a banquet, he was accidentally fed a portion of rancid food. Rather than throw away the food, Caesar took a large helping of the rotted cuisine and ate it happily.

Born in the year 100 BCE, Julius Ceasar's father died when he was only fifteen years old, just after the end of the Social War. While still young, Caesar married Cornelia, the daughter of the Consul Cinna, who named him to the post of Flamen Dialis, which was an obscure religious post, but it gave Caesar his first taste of power. Rituals, pomp and circumstance surrounded the Flamen Dialis. They were not allowed to even look at anything that was dead and could not wear clothing with any knots in it. As Flamen Dialis, Caesar was effectively prohibited from most other political offices. Most of the people who gained great power in Rome were from wealthy, powerful families with many political connections. Perhaps it was thought by Cinna that if he gave the young Caesar this one post, he would be satisfied with this post and lacking the money usually required to win higher office, would be limited in the power he would gain. Instead, this only fueled his passion for politics and his quest for power.

Sulla retook power from Cinna and annulled all the acts passed by the former counsel, including Caesar's post and his marriage to Cornelia. The young upstart refused the order to divorce his wife and even though he was sick with a fever at the time, he fled Rome, in fear for his life. He was finally caught, but due to his family lineage, he was saved from execution, despite the warning of Sulla that Caesar would grow to be more dangerous to Rome than anything that Rome had yet faced. History would later prove Sulla ~~to have been~~ correct.

Caesar then served in Asia Minor and visited with King Nicomedes in Bithynia (in modern-day Turkey). He spent so long with Nicomedes that Caesar was accused of having a homosexual relationship with the king. By this time, Caesar had become a great orator and even his enemies had to admit that he was among the smoothest talking people and that he gave the greatest speeches, in the Republic.

On his way home by boat from Asia Minor, Caesar was captured by some of the pirates which had become so prevalent throughout the Mediterranean. They ransomed him for fifty talents[*], which Caesar had delivered to the pirates in exchange for his freedom. While a prisoner aboard their ship, he supposedly told the pirates that after they released him, he would have them all executed. After his release, he did just that.

Back in Rome, he was named pontifex (a member of the highest council of priests in Rome) in 74 BCE and he became quaestor (which made him responsible for the finances and administration of the military and government) in 69 BCE.

The very same year, both his aunt Julia and his wife Cornelia died, leaving

[*] Both in Greece and Rome, the value of a talent was between ten and thirty kg (23-70 lb) of gold. At today's rate, that would be up to nearly half a million dollars.

Caesar heartbroken. Of course, even this moment of mourning for deceased family members did not stop Caesar from making a political spectacle of their funerals. Despite the fact that women were not usually given large funerals, Caesar made their funerals a great escapade. He declared his aunt had been descended from Gods and kings (the king part was partially true if you went back far enough in the Julian line). What was left unspoken was his insinuation that if Julia had been descended from kings and Gods, then he was, as well.

As his next wife, Caesar took Pompeia, the granddaughter of Sulla, as his bride. This was mainly a political move, to appease the Sullans, one of Rome's most powerful families. He also became a protégé of Crassus during this time, another shrewd political move, as Crassus later financed Caesar's successful campaign for aedile in 65 BCE. As aedile, Caesar was in charge of public works and the water system, along with the public games and the grain supply for the city.

The office of Pontifex Maximus was considered one of the most honorable offices in Rome, usually filled by former consuls and censors. When the year 63 BCE came around, Caesar stood for the office against the person most people favored, Metellus Pius. He left from the home of his mother the morning of the election with the words "You will next see me as Pontifex Maximus, or not at all!"[1] He won the election, possibly through extensive bribery.

Caesar became propraetor in 62 BCE as the Catiline Conspiracy was taking shape and he was assigned to a unit in Hispaniae (modern day Spain). However, due to his massive debts, his creditors asked for and received, an injunction that forbade Caesar from leaving the city. Crassus guaranteed the payment of Caesar's debts and he left for Hispaniae, where he became notorious for destroying cities that had already surrendered to his forces. After a year of war in Hispaniae, Caesar had collected enough wealth through the spoils of war to pay off his debts.

Now Caesar wished to run for the office of Triumphator, however Roman law forbade any general-in-arms from entering the city. Caesar asked the Roman Senate to grant him permission to campaign *in absentia* and this was denied, largely due to the influence of Cato the Younger. Caesar abandoned his post and joined Crassus and Pompey in the formation of the First Triumvirate, in order to stand up to anything the Senate might do to check the rapidly growing power of Caesar. Between the three men, they controlled the offices of consul and tribune (this office had the power to veto any legislation the Senate passed), they had massive military forces and a nearly limitless supply of funds. The First Triumvirate was never an official legal body, but between the three men, they held reign over Roman lawmaking, supplies and the economy.

The Triumvirate ruled that land not used in Italy would be given to the men who served under Pompey. Then, Caesar decreed that he would be given command of forces in Gaul, where he was responsible for the deaths of over a million Gauls and

---

[1] Matyszak, Philip. "Chronicle of the Roman Republic." 2003.

where he sold over another million of their people into slavery over the course of ten years. This genocide also came at the price of the lives of hundreds of thousands of Romans. However, ever the self-promoter, Caesar wrote up his military campaign there in *The Gallic War*. This piece, written by Caesar in the third person, gives great insight into the mind of Caesar, even as it glosses over his human failings, military ruthlessness and lust for power.

During his military campaigns, he developed what became known as the *Caesar Cipher*. This simple substitution cipher code was used to encode messages between Caesar and his generals. This was a simple code, in where each letter was substituted for another letter, usually with a shift of three letters. For instance, the letter "A" would be written as "D", "B" as "E", "C" as "F" and so on*. The name "Caesar," for instance, would be written as "Fdhvdu." In today's world, that would be an astonishingly simple code and would be broken by an attacker within minutes, but we must remember that many people at the time were illiterate and just normal writing would have been safe for keeping a good portion of the people from reading the message.

Caesar daughter Julia, who was also the wife of Pompey, died in 54 BCE, but bad feelings were already growing between Caesar and Pompey. Pompey had been infatuated with his bride to the point where he had begun to ignore politics and warfare in order to spend more time with his young wife. Pompey also expanded his private life, by commissioning Pompey's Forum, the first permanent shopping center/entertainment center in Rome. Caesar despised Pompey for his waning interest in military and political power and this would grow to have direct consequences for the destruction of the Great Library at Alexandria.

The following year, Crassus was killed. During the revolt of Spartacus in the Third Servile War, Crassus had bought himself the ultimate male fantasy gift; his own army. He used this army when he tried to trap Spartacus and his followers on the toe of the Italian mainland and failed. He may have been able to buy himself an army, but he was never a great military leader. Displeased with the performance of several of the units in his army for failing to catch Spartacus the first time, he ordered the execution of every tenth man in the units whom he deemed to have been responsible for his loss. It was not unheard of to execute every tenth man in such units. In fact, it is from this practice that we get the modern word *decimate*, since the Latin word for ten is *decem.*

In 53 BCE, Crassus led his army to war against the Parthians and lost badly. Of the 6,000 men under his command, 5,500 were killed outright and the rest, including Crassus, were taken prisoner. Crassus was executed and molten gold was poured into his head to exemplify his great greed.

With the power vacuum caused by the death of Crassus, Caesar and Pompey began a fight that would grow into a Roman Civil War for control of the Republic.†

---

* Octavian also used a similar cipher, but he used a shift of one letter; "B" for "A", "C" for "B", etc.

† Incidentally, this is the point in the historical timeline that the HBO series "Rome" begins.

Caesar, still assigned in Gaul, was quickly losing what support he had secured in the Roman Senate and they ordered Caesar to abandon Gaul and relinquish his power to Pompey. The Senate had also refused the candidacy of Caesar for consul to rule in 48 BCE. They intended to arrest Caesar on his return to Rome, for now his political rivals were banding together in order to attenuate the power of the Triumvirate.

Caesar refused to abandon his post and instead took his Thirteenth Legion out of Gaul, crossing the Rubicon River into the Roman Empire itself on January 10, 49 BCE*. There was no military engagement here, but the very act of bringing an army into the Roman homeland was itself illegal and Caesar was now leading a fully armed rebellion against Rome. The Senate quickly moved to give command of the Roman legions to Pompey and the Roman Civil War had begun. Caesar was no longer a rouge politician. He was now in open rebellion against the Roman Empire and an enemy of the state.

The Roman legions adopted a new marching song, which spoke of Caesar coming into Rome, to take over the republic:

Gaul was brought to shame by Caesar; by King Nicomedes, he.
Here comes Caesar, wreathed in triumph for his Gallic victory!
Nicomedes wears no laurels – though the greatest of the three.
Home we bring our bald whoremonger; Romans lock your wives away!
All the bags of gold you lent him, his Gallic tarts received as pay.[2]

City after city fell to Caesar, many of them without a fight and it looked like Caesar was about to gain control of all of northern Italy. In just a few weeks, Pompey and his allies in the Senate fled the city, declaring anyone who stayed to be an ally of Caesar and an enemy of the state. In their great haste to leave the city they left the Roman treasury behind, unguarded, under the Temple of Saturn (Chronos) in Rome.

Once in Rome, early one morning, Caesar and his bodyguard went to the temple to impound the treasury. A young Roman guard by the name of Marcellus had the audacity to tell Caesar that the money rightfully belonged to the government of Pompey. Caesar then explained to the guard that it would be even easier for Caesar to kill the guard, than to threaten to do so. Marcellus stepped aside and in one day, Caesar had a greater amount of money for his forces than Pompey.

However, Caesar was not always ruthless military leader (despite his exploits in Gaul). Far from it, he was known to have pardoned entire cities and armies which had fought against him in the civil war, even to the extent of telling the citizens that if they

---

* Although used less often in modern times, the phrase "To cross the Rubicon" is still meant to mean taking a irrevocable action which shifts strategy greatly.

[2] From Suetonius, in the "Life of Caesar." Quoted by Philip Matyszak in "Chronicles of the Roman Republic." 2003.

wished to go back and fight for Pompey, that was their determination to make. This won him many admirers and converts to his side, but oddly enough, of all the three dozen or so people who would conspire in the assassination of Caesar, all but one had been so pardoned by Caesar during this time.

Caesar still did not have much of a navy at his command at this point (unlike Pompey) and so he took control of the western part of Europe and northern Africa, leaving Pompey in firm control of the eastern Mediterranean. Pompey and his troops solidified their positions in Greece and waited for the attack they knew would come from Caesar.

Caesar followed Pompey to Greece, where his troops were so badly supplied that they were forced, at one point, to eat bread made from grass. The two leaders met at the Battle of Pharsalus, in southern Thessaly, on 9 August, 48 BCE. This was a battle of heavy infantry on both sides. When it was over, Pompey was defeated and left for Egypt. Caesar followed him there and set up a palace in Alexandria. It was there where he would meet the most famous queen of the ancient world.

The Seleucid Dynasty, bearing the name of the Ptolemys (who had founded the Great Library), continued to crumble in the first years of the first century BCE. The last of the Ptolemys and the most famous of all, was Cleopatra VII, the infamous queen, renowned throughout the world and for over two thousand years of history.

Born in 69 BCE, Cleopatra grew up in Alexandria, which, at that time, was the most populous city in the world. Despite being born in Egypt, Cleopatra did not have a drop of Egyptian blood in her veins. Ironically, the woman who would go on to become the most famous queen of Egypt, as well as the land's last pharaoh, was pureblooded Greek.

For two centuries, Egypt had been losing her colonies and Rome became ever stronger. Her father had capitulated to Rome on many occasions and was paying them a hefty tribute. Egypt was deeply in debt to Rome and her father, Ptolemy Auletes (Ptolemy XII), was seen as a Roman puppet and a sell-out to the world's only superpower at the time.

Upon the resignation of Auletes due to his failing health, Cleopatra took the throne and she was forced to marry her twelve-year-old brother, Ptolemy XIII. After all, an Egyptian queen at that time could not rule without a king and any spouse she chose had to be of royal blood.

At just twelve years old, the young co-regent Ptolemy XIII ordered the execution of Pompey. On his arrival in Alexandria, Pompey was taken on a trip in a small boat with two associates of his from his early days in the military. While at sea, they stabbed Pompey and decapitated him, leaving his naked body unceremoniously on a shore. With two of the three members of the First Triumvirate now dead, Caesar was the unquestioned sole ruler of the Roman Republic.* Caesar was visibly upset

---

* Although the Republic still existed in name, it was quickly falling into a long-term dictatorship at this time. Ironically, the Roman Empire was not much more of an empire that the Roman Republic

when the head and ring of Pompey were brought to him and he began to cry, calling the act treachery. It is likely that Caesar wanted Pompey alive, so that they could once again become allies and rule Rome together.

Cleopatra is renowned today as a great beauty, but it is more likely that she had a very ordinary face and certainly had a large nose, which was a family trait amongst the Ptolemys. However, it was her charms that would win both Caesar and Marc Antony to her side. It is said that Cleopatra spoke nine different languages, although this may be an exaggeration. What is known is that Cleopatra VII was the first of her family to speak the Egyptian language, which helped endear her to her people.

Caesar, now fully in charge of Rome, had come to Alexandria with his legions to collect the debt owed to him by Egypt, as well as to chase Pompey. Anti-Roman riots broke out on the street and Cleopatra went to visit Caesar in his Alexandrian palace.

Caesar, a well-known womanizer, was fifty-two at the time and Cleopatra, twenty-one. She executed her entrance perfectly, having herself wrapped up in a carpet and rolled out before Caesar. After an evening of drinking between the two, Caesar agreed to back her in a quest for the throne, in order to place Cleopatra in sole control in Egypt.

---

had previously been. Almost all the territorial gains that we associate with the Roman Empire were secured during the period of the Republic.

An Emperor Falls, an Empire Rises 48 BCE – 20 CE

A portrait of Octavian, after he rose to emperor and became known as Augustus.

Public domain photo

Ptolemy XIII saw that Caesar and Cleopatra had taken both a personal and a military partnership with each other and he declared that this act and the backing of Cleopatra for sole ruler was treason against Egypt. His 20,000 troops surrounded Caesar and Cleopatra in Alexandria. During the battle, Caesar set the enemy ships in the harbor on fire and some ancient historians tell that this fire spread to destroy several thousand scrolls that belonged to the library.

Whether this event occurred, or how great the damage was to the library is a point of great contention. Caesar himself wrote of his exploits here in "The Alexandrian War" and makes no mention of such a fire. Even critics of Caesar from his own time make no mention of it either. The first time the story of this fire is put into words is one hundred years later, by Seneca. Depending on whose story you are reading, somewhere between 40,000 and 700,000 scrolls were destroyed by this fire.

Plutarch (who was the first to even mention Caesar by name, about 117 CE) even states that the fire destroyed the library. However, this seems unlikely, as the library was located well within the city and a fire in the harbor should never have been able to spread that far. It is possible that the fire may have destroyed some warehouses near the port, or perhaps there were scrolls packed on the docks, waiting for shipment somewhere else. We will never know what happened here, but if there were a significant number of scrolls destroyed in this fire, than this represents the first of many disasters that would strike the ancient library.

Just a few weeks after Cleopatra and her brother reunite in Alexandria, Ptolemy's supporters staged an attempted coup and civil war broke out in Egypt. Soon, the rebels were defeated and Ptolemy XIII was found dead in the Nile, evidently killed while attempting to flee.

Cleopatra then found that she was pregnant, likely by Caesar. She gave birth to a son, whom she named Caesarion. Meanwhile, Caesar was holding a celebration in Rome in honor of his victories, including a decisive win over King Pharnaces of Pontus. It was at this celebration that Caesar uttered his immortal words "Vini, Vidi, Vici", meaning "I came, I saw, I conquered."

Also at this time, Caesar made a change that still resonates to our modern day. The Roman calendar, which was based on twelve moon cycles per year, had become highly inaccurate due to the fact that there are less than twelve moon cycles per year. In fact, it was so bad that January had crept into the autumn. Caesar declared that the year 46 BCE would be extended and that the year 45 BCE would begin on a declared day to be deemed 1 January and that each year would have 365 days, except for every fourth year, which would be a leap year containing 366 days. The Senate then voted to rename the month of Quintilis (the fifth month, as quinque is the Latin word for five) to July, in honor of Julius.* At this time, the month of February had twenty-nine days, except for leap years, when it would have 30 days. March would have thirty-one and so on. This would change.

Cleopatra VII with her son, Caesarion.
The son of Cleopatra and Caesar lived to be seventeen years old.

Public domain photo

Cleopatra sailed to Rome, with her son in tow and when she arrived in the capital city, she found that Caesar had erected a statue to her in the temple to Venus in Rome.

However, this was a different Rome than it had been just a few years before and the rumors spread wildly that Caesar was planning on abolishing the Senate and appointing himself sole leader of the Roman Empire. There certainly was some truth behind those rumors, as this is almost exactly what Julius Caesar would go on to do.

At the beginning of the year 44 BCE, Julius Caesar declared himself "dictator for the restoration of the commonwealth." This was interpreted by many to mean "dictator until I resign or die." The Roman people were used to dictators; in times of crisis (usually war), they would *elect* a dictator. However, the terms of these elected dictators were only for a six-month period. This was entirely different, as there was no sight in end to the absolute reign of Caesar.

Now, nothing was left for Caesar on the political front, as he had achieved supreme rule in the greatest empire in the world. He began to dress in the robes of kings and take on all the trappings and benefits of such an office. However, he also attempted to endear himself to the people of Rome by refusing the title of king. At one point, he even turned down the Senate's offer to exempt him from all laws and to allow him legal access to any woman in the Empire he chose. However, many people were uncomfortable with the overthrow of the Republic and conspiracy was brewing.

---

* One may wonder why Quintilis was called the fifth month when it is the seventh month in our modern calendar. This is because the old Roman calendar had ten months until the addition of the months of January and February at the beginning of the seventh century BCE. The calendar until Caesar began in March, so that the last four months (September, October, November and December) actually used to match up with the Latin numbers for seven through ten (septem, octo, novem, decem). Since February under this format was the last month, this is where they placed the extra day for leap years.

Marcus Junius Brutus; The most famous of Caesar's assassins.

Public domain photo

On the ides of March in 44 BCE, a group of Roman senators, led by Marcus Brutus and Cassius, murdered Caesar. Marcus Brutus was descended from Junius Brutus, who had lead the rebellion against the Tarquins centuries before, overthrowing the earlier system of elected kings, helping to form the Republic. Marcus Brutus almost certainly did what he believed to be right, assassinating the man who had overthrown the Republic. Over forty senators were involved in this conspiracy, most of them likely used to hold back the other Senators who might otherwise attempt to save the life of Caesar. The man who reworked the calendar and had a month named after him only lived to see one month of July, as he was assassinated the following March.

Just before the assassination, Caesar had been warned to "beware the Ides of March" from a soothsayer and that morning, his wife awoke from a terrible dream that her husband had been murdered. Due to the fires in Rome, the Senate was not meeting at the time at the Senate house and Caesar went to meet them a few hundred meters (yards) away from where they normally would have met. On the way into the building, the soothsayer was again seen on the steps. When Caesar asked him what had happened to his prophecy of the Ides of March, Caesar was told that "The Ides are not yet over."

Strangely, Caesar took not a single bodyguard with him for his trip to meet the Senate and as he was carried along the streets, many of the people of Rome jeered him as he strode toward his meeting with fate. Several of these people handed him notes, most of which were taken by aides and never seen by Caesar. However, there was one note that Caesar held on to that was found on his person after death. This was a message warning Caesar that he would be assassinated that day, at that meeting. Caesar must have read the note and so likely knew that there would be an assassination attempt on him within minutes. Why he did nothing to stop his death is a mystery. It is possible that Caesar did not believe the note, or that he may have wanted to die, either in order to finally rid himself of his ills, or because a magnificent death would be the perfect end to a man who went from massive debt to the leader of the world's only superpower.

The assassination of Caesar left Rome divided between Marc Antony, a trusted aide to Caesar, and Octavian. Cleopatra quickly fled back to Egypt. Soon after she returned, her second brother/husband, Ptolemy XIV, was murdered, possibly with the assistance of Cleopatra. Being forced to name a family member as co-regent, she named Caesarion as co-ruler of Egypt.

Back in Rome, Octavian took charge of Rome in the form of the Second Triumvirate, with Marc Antony and Marcus Aemilius Lepidus in 43 BCE. However, this was a legal formal body which was decreed by the Senate and given sweeping powers to do as they wished. Like the First Triumvirate, the personalities would clash, but unlike earlier, this would finally destroy the alliance.

A coin from 39 BCE, showing the face of Marc Antony.

Photo courtesy Classical Numismatic Group

Despite previous events, they associated themselves with the name of the fallen emperor, creating support for what was, effectively, a military junta by evoking the name of Julius Caesar. They killed three hundred senators as well as Cicero, whose head and hands were hung in the forum. One of the first acts once their power was firmly established was to raise taxes on wealthy women (none of whom were allowed to vote). One woman, named Hortensia, gave a speech in the forum against the higher taxes on women still without suffrage, where she declared a phrase which would later be spoken by the founding fathers of America: "No taxation without representation."

Lepidus had been a military leader, whose father (not surprisingly of the same name) had been defeated by Pompey in northern Italy and forced to flee the peninsula years before. Lepidus had been elected to office several time under Caesar; he made praetor in 49 BCE and was consul under Caesar three years later.

He had the foresight to support Antony during the Civil War and he had been rewarded with a seat on the Second Triumvirate. However, this would not last long. The Battle of Philippi occurred in 42 BCE, which pitted Antony and Octavian against Brutus. After this battle Lepidus was granted control of Africa. He then conquered Sicily in 36 BCE and Octavian grew suspicious of the goals of Lepidus and did not believe that Lepidus would share control of Sicily. Octavian had Lepidus stripped of his powers and title, except for the title of Pontifex Maximus.

Cleopatra had re-built the economy of Egypt and now the Egyptians had a good deal of wealth; a commodity much desired by Marc Antony. Antony was a large, robust man, on his third marriage. A womanizer like Caesar, he differed from the previous leader in the dubious fact that he was also a drunkard, with little self-control. But, his propaganda machine spun this beautifully, portraying Antony as the new Dionysus, the God of (among other things), wine.

Marc Antony was planning to invade Parthia (now part of Iran) and needed the support, military and financial, that Cleopatra was able to supply. The two leaders and lovers began to portray themselves, in statues and etchings, as the God and Goddess Dionysus and Isis - a perfect combination of Roman and Egyptian deities. Their dream was to unite Egypt with Rome, which would allow them to control almost the entire Mediterranean, regaining the power that Egypt had been given after the death of Alexander the Great. During negotiations between the two, Cleopatra demanded that the former empire which had been ruled by Ptolemy I be restored and Marc Antony granted her request.

Two Syrian coins from 32-31 BCE, depicting Cleopatra and Antony on the face of the coins.

Photo courtesy Classical Numismatic Group

Marc Antony later invaded Parthia and was defeated. At a dinner soon afterwards, Antony and Cleopatra showed up dressed in full costumes, as Dionysus and Isis. There, Marc Antony officially declared Caesarion to be the son of Caesar, growing the bonds between Rome and Egypt. A new coin was produced in the Roman Empire at this time, depicting Cleopatra on the face, in a matronly, older pose, designed to create a greater illusion of the connection between Cleopatra and Isis in the minds of the Romans.

When war finally broke out between Cleopatra and Marc Antony on one side and Octavian on the other in 31 BCE, the pair sailed with 500 ships and over 75,000 men. This was the largest fleet seen since Alexander the great, nearly 300 years earlier. The two forces, which pitted Roman against Roman, met at Actium, on the western coast of Greece, on 2 September, 31 BCE. This would be the last great sea battle in the history of Rome and most of the ships were later retired by Octavian.

The Battle of Actium, depicted in a painting by Lorenzo A. Castro in 1672.

Public domain photo

While the battle raged, Marc Antony left and headed back towards the port of Alexandria, with Cleopatra's ship (containing the Egyptian treasury) following close behind. Although this is usually seen as a retreat, it is possible that Marc Antony had planned this and expected his other ships to follow (which they did not). The great naval force they had assembled began to defect, ship by ship, to Octavian.

Marc Antony and Cleopatra made it back to Alexandria, but their time was numbered. The people of the city began to refer to them as "Those who are going to die together." As Octavian closed his forces around Alexandria, Cleopatra offered to abdicate her throne, if her children could rule in her stead. Antony offered his life in return for Octavian sparing the life of Cleopatra. That night, the city's population heard horns leaving Alexandria and word quickly spread that the sound was symbolic of Dionysus leaving the city.

The next morning, in a scene that would be rewritten 1600 years later in Romeo and Juliet, Marc Antony heard the (false) news that Cleopatra was dead. In mourning, he committed suicide. Cleopatra knew Octavian's army was in Alexandria and that, if she were caught, she would be paraded as a trophy around the streets of Rome. Unable to live with that fate, she allowed herself to be bitten by a poisonous snake, likely an Egyptian cobra.* Two puncture wounds were found later on her body.

As she lay dying, Octavian entered the room. Her last words were to him, requesting that she be buried next to Marc Antony. She was only thirty-eight.

After the death of Cleopatra VII, Egypt became part of the Roman Empire and Octavian murdered Caesarian, in order that the Ptolemaic dynasty would finally be ended. When the vast treasures of Alexandria were brought back to Rome, interest rates in the Roman capital dropped twenty-five percent almost instantly.

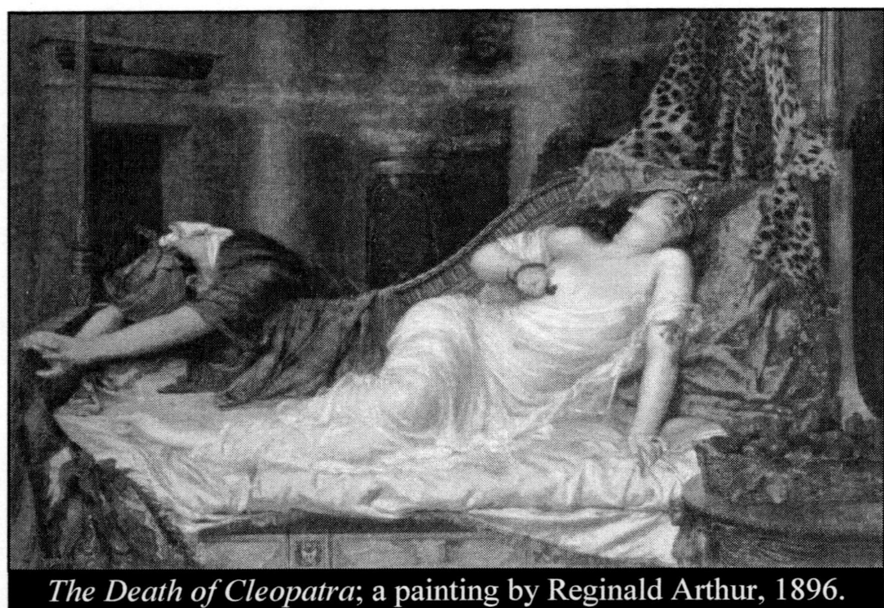

The Death of Cleopatra; a painting by Reginald Arthur, 1896.

Public domain photo

Octavian ruled Rome for the next forty-five years, as the Emperor Augustus, until the time of Christ. During his reign, the Roman Senate further

* The Egyptian cobra is a type of asp. A legend at the time in Egypt said that anyone who died of a cobra bite was granted immortality in the afterlife. This may explain the choice of a cobra as a method of suicide. There is also reason to believe that she may have drunk poison instead of being bitten by a snake. Then again, she may have done both.

changed the calendar, in order to please the new emperor. Since they had named a month after Caesar, they decided to name a month after Augustus, as well. Thus, the month of Sextillus (*sex* is the Latin word for six) became Augustus, or August. Never ones to overlook potential insults to those in power, the Senate realized that August only had thirty days, unlike July, which had thirty-one days. Therefore, they added one extra day to August, which is why July and August are the only consecutive months on the calendar that both have the same number of days. The extra day was taken from February, changing the length of that month to twenty-eight days, except for leap years when it lasts for twenty-nine days. Now they had a calendar that had three months in a row (July, August and September) that each lasted thirty-one days. In order to avoid this, they changed the order of the days starting in September, so that September had thirty days, October lasted thirty-one days, etc.

Under Augustus, the transformation of Rome from a republic into an empire was complete. In many ways, Augustus would be one of the great Emperors of Rome. His reign was long and stable, the newly-renamed Roman Empire had finished decades of civil war and rebellion and was mostly at peace.

Twelve years after the battle of Actium, in 19 BCE, the Roman poet Virgil wrote his masterwork "The Aeneid," an epic poem of war. The work was nearly complete when Virgil died and the author gave an order on his deathbed that the epic poem should be destroyed. However, this order was overruled by Augustus, who directed that the poem should be edited and printed.

This was also a time when the infamous Roman toga began to fall out of favor. Most people began to wear just a tunic without a toga, especially in the summer. Many of the people found time and money for luxuries and recreation and began to vacation more often as living conditions worsened for the vast majority of the people of Rome; the patricians became richer as the plebeians became poorer.

Towards the end of the reign of Augustus, he became concerned with maintaining a "traditional" Roman lifestyle for the people of the Empire, including fighting against the coming style in dress by advocating the wearing of togas. This was an uphill battle he would lose. Augustus, the first true emperor of Rome, would die just a few years before the birth of our next great scientist, Heron of Alexandria.

Heron of Alexandria c. 20 – 62 CE

Heron of Alexandria; The man who nearly started an industrial revolution in Alexandria

Photo believed to be public domain

In 14 CE, just a few years before the birth of Heron of Alexandria, Rome's first emperor after Julius Caesar, Augustus (the former Octavian), died. He left the Empire to Tiberius, who was an exceptionally unpopular choice. When the news spread to the forces headed by Germanicus (the husband of Augustus' granddaughter, Agrippina, later known as Agrippina the Elder), they readied to rebel against Tiberius, in order to put Germanicus in power as emperor. Germanicus sent his son, Gaius, along with his sister Agrippina (the Younger) away, in order to keep them from danger.

Gaius had been born in Antium (currently Anzio, Italy), the son of Germanicus, a military leader, five-time quaestor and consul. Gaius (whose full name was Gaius Julius Caesar Germanicus) was brought up by his father's side in the military, as Hannibal had been. However, unlike Hannibal, young Gaius was made the mascot of his father's units when he was a toddler. He was made to dress up like a little soldier, including little soldier's boots. It is from the word *caliga* and the infix (modifier) *ul* (little) that Gaius got his nickname, Caligula. He never liked this nickname, but he did not care for the name Gaius any better.

The troops, believing that their mascot no longer being with them was a bad omen, laid down their plans for rebellion against Tiberius. Agrippina and the young Caligula returned to their father's side.

Tiberius continued to rule the Empire from then until 37 CE. Despite a general dislike of Germanicus, Tiberius adopted him as his son soon after becoming emperor. When death came to Tiberius in 37 CE, he in turn, left control of the empire to his grandson, Tiberius Gemellusa and his grandnephew, Caligula, who was now in his mid-twenties.

On 18 March, 37 CE, the Roman Senate voided the will of Tiberius and Caligula was in sole command of the Roman Empire. Having never been on the public spotlight before, Caligula was extremely nervous and uncomfortable with the attention he received.

At first a reasonable, sane man, he helped those who had been harmed by the draconian tax system of the Empire and ended a series treason trials, which had begun under Tiberius. Not only the son of Germanicus, he was also a descendant of Caesar and Octavian and the great-grandson of Marc Antony.

Caligula contracted a high fever about six months into his reign, possibly encephalitis (an inflammation of the brain). His father-in-law died soon after Caligula recovered and this also hit Caligula deeply. He was never the same afterwards and went on a killing spree that is still known to us today. He murdered most of his family including Gemellusa, wasted his entire family fortune on entertainment and events, ordered one public works project after another that Rome could never afford and had people tortured and murdered while he ate, as dinner entertainment.

A bust of Caligula in the Louvre in Paris

GNU public license photo

Otherwise, he could be extremely suggestible and willing to do as others told him to do. He went on to have himself named a God and ordered the construction of many temples to himself, as well as regular sacrifices in order to please him. When required to name a Consul for Rome, which was an appointed position in the imperial system, Caligula assigned his favorite horse, Incitatus, to the position.

Incitatus lived in a marble manger and ate hay mixed with gold flakes inside a stable of marble. Wearing a collar bedecked in precious jewels and wearing purple blankets (the color of kings), "he" would invite heads of state and other VIPs to meet with him.

In 41 CE, members of his own bodyguards, the Praetorian Guard, named Cassius Chaerea and Cornelius Sabinus, murdered Caligula by jumping him in a hallway. They then brutally murdered his wife and infant daughter. Caligula was finished and Claudius would take his place as emperor.

Heron was entering his twenties at this time. His name can also be written as Hero and it was another common name at this time in history. Eighteen Greek writers are known from this time that were all named Hero or Heron.* He walked the floors of the Great Library about the same time that Caligula was murdered and that the city of Lugdunum (present day Lyon, France) was founded by the Romans in Gaul. This city was a trading city, run by common people and freed slaves who prospered over the tremendous wealth moving through the city, controlling the trade in wine, oil, corn and lumber for the Empire, as well as being the manufacturing center for nearly all the goods consumed by the people of Gaul, Germania and Britain. Without any history of elite ancestors, these merchants gained great power from their control of trade throughout the Roman world.

Heron of Alexandria appears to have been of Egyptian birth and to have written at least thirteen books on the subjects of mechanics, physics and mathematics. Many of these appear to be textbooks written by Heron for his classes and others may have been collected later from his lecture notes from a diverse range of classes on physics,

---

* For those of you keeping score, that's Hero/Heron 18, Ptolemy 12 (and one more to come), Cleopatra 7.

mathematics, mechanics, pneumatics and more.

In his book "On Dioptra," Heron gives his theories on astronomy and surveying. He shows how the distance between Rome and Alexandria could be calculated from the difference in time between when the two cities observed a lunar eclipse. He also wrote works on engineering, weapons of war, hydraulics, measurement, optics and other fields of study. He may have also used some of the devices he invented to teach his students about physics. He had artificial animals that would drink water when it was offered to it, puppets that moved of their own accord when a fire was built underneath the figure and even a mechanical singing bird to show the power of steam.

Heron is best known as a mathematician, for his work on the measurement of geometric forms and shapes. He also developed a system for the estimation of the square and cube roots of numbers that are not perfect squares or cubes. He is often credited with a formula for the area of triangles, but this was likely known before his time. He believed that the field of mechanics could be dived into two parts: theoretical and manual. The theoretical part under this distinction was consisted of physics, astronomy and mathematics. The subjects that made up his manual category were architecture, carpentry, blacksmithing, painting and other practical endeavors.

Alexandria in the first century CE was the most cosmopolitan city in the world and the inventions of Heron nearly gave Alexandria a jump of almost 2,000 years in creature comforts and technology. Had the people of Alexandria been widely exposed to these mechanical miracles, it is possible that what we call history today would be far different from what transpired.

Nevertheless, Heron was not the only scientist contributing to science in Alexandria at that time. Mary Hebraea was a chemist working at the Great Library at this time and she invented a chemical found in science labs everywhere today, from junior high schools to advanced research labs. Habraea was the first person to discover hydrochloric acid.

There was another woman who had a cathartic effect on Rome at this time, although she too remains almost entirely unknown to the world. In 43 CE, two years after the assassination of Caligula, the Roman Emperor Claudius ordered the invasion of the island that would become known as Britain. The Roman rule here was harsh and unforgiving. For twenty years, their despotic rule taxed the people of the island into submission and Romans treated the people of the island like a defeated enemy.

They would regret their arrogance when they subjugated the people of the Iceni, who lived on the easternmost point of the British mainland. When the king of the Iceni, Prasutagus, died in 60 CE, the Romans thought they saw a chance to bring these people under their control – The king was dead, but his wife Boudica still lived. Boudica was a strong, tall woman with a powerful voice, a fierce look in her eye and flowing masses of reddish-brown hair. She is said to have dressed in colorful tunics, over which she wore a loose, open cloak, fastened with a brooch and from her neck

Baudica of the Iceni. From a wall portrait at the Colchester, England town hall.

Photo believed to be public domain.

hung a golden medallion. When the Romans came for Baudica, she was beaten and whipped, many of her people were enslaved, including her family and her two daughters were raped.

This was more than enough for the queen of the Iceni and she put together an army of her own to fight the Romans – other armies in the area soon allied with her against the imperial forces. Boudica destroyed the Roman colony of Camulodunum, which is modern day Colchester. She also laid siege to Londonium and Veralamium (modern day London and St. Albans). During these battles, she avenged the destruction of her people and the treatment of herself and her family by killing over 70,000 Roman soldiers.

The Roman governor, Suetonius Paullinus, came back from his battle on the island of Mona, where he was attempting to subdue the Druids, to fight this mighty woman warrior. He got to Londonium and found that the requested reinforcements had not arrived and quickly abandoned the city, leaving the most populous city on the island to Boudica. Not only had the Romans lost a great deal of territory, but they had lost it to a woman. Paullinus was enraged and decided to meet her army with his own, which numbered about 10,000 men.

However outnumbered they were, Paullinus had the advantage of picking the spot for the battle and the area he chose (near Londonium) gave him a tactical advantage. By now, the army of Boudica had swelled to tremendous numbers, perhaps as high as 230,000, although that number seems unlikely. The larger army, however, was also overconfident in their assessment of the force they could bring against Paullinus. Certain of victory, the men in Boudica's army brought their wives to the scene of the battle to witness their "inevitable" triumph, as the wives sat in wagons stationed around the battlefield.

The battle began and Boudica drove through the crowd with her daughters in a chariot, shouting encouragement and giving orders to the assembled masses. Meanwhile, the Romans kept up a steady, orderly march through the hail of arrows and when they got within a javelin's throw of Boudica's forces, they launched into the crowd with a volley of javelins and then ran into the crowd in a wedge formation, separating and dividing the forces of Boudica.

At first, the chariots of Boudica broke through the range of archers under Paullinus' command. But the chariot drivers had no breastplates and were soon driven back into the melee. The army fighting the Romans tried to flee but were blocked by the wagons holding the wives of the soldiers, as well as the corpses of animals

A woodcut of Heron's aelopile. This device, considered a toy, is a critical step to developing a steam engine.

Public domain photo

sacrificed before the battle. From here, it was a rout and the Romans not only slaughtered as many fighters as they could, but they also massacred the wives that were in attendance. Perhaps 80,000 people were killed that day at a cost of only 400 dead Roman soldiers and another 400 injured. Boudica herself was so distraught with this loss that she committed suicide by poison at the scene of the battle.

Heron's many inventions earned him the title *The Machine Man*. He is known to have developed the world's first vending machines, automatic doors, self-controlled stage props, screw-cutting machines, scenery and sound effects, along with the first ever machine gun and mechanized con machine. He certainly developed the first known form of programming in his mechanized devices and he even invented the world's first true computer programming.*

He is also known to have built a rotary steam turbine, known as an aelopile. This was, very nearly, a true steam engine. The aelopile was crafted in the form of a metal sphere, which was held within a bracket above a large closed pot. It had two "L"-shaped tubes coming off it in such a way that when the water inside the pot was heated, the steam coming up the brackets and out of the tubes would cause the device to spin very quickly. It was the fastest rotating object in the ancient world.

More intriguing, had Heron hooked the aelopile up in such a fashion that it could do work, for instance grind corn, or mill textiles, or drive a wheel on a chariot, it might have set off an industrial revolution 1,600 years before one finally took root. There are may reasons that Heron may not have made this connection that would have dramatically changed what we call history. Perhaps he never saw the potential of what appeared to be just an idle curiosity: a spinning sphere. Perhaps the fact that plentiful supplies of slavery and cattle were available in Greece, Rome and Egypt made laborsaving devices impractical. After all, the so-called "Pompeian mill" for grinding

* Although the earlier Antikythera was an analogue computer, it could only produce a real-time calculation. This is just as if you spun the minute hand of an old-fashioned mechanical wristwatch a certain number of minutes and calculated from the position of the hour hand how many hours had passed. The advance of Heron would allow his devices to execute a *series* of tasks, in a specified order and for the desired amount of time.

corn and grains caught on very fast in the second century BCE because it was driven by cattle (usually donkeys) and was very inexpensive to obtain and maintain. Water-driven mills were much more expensive, usually requiring that an aqueduct be built to bring a steady stream of water to the mill. Perhaps Heron saw the potential of this new source of power, but no one was willing to pay the money to develop the device. We will possibly never know why Heron came so close to such a massive paradigm change in society and never took that last half step to mass production.

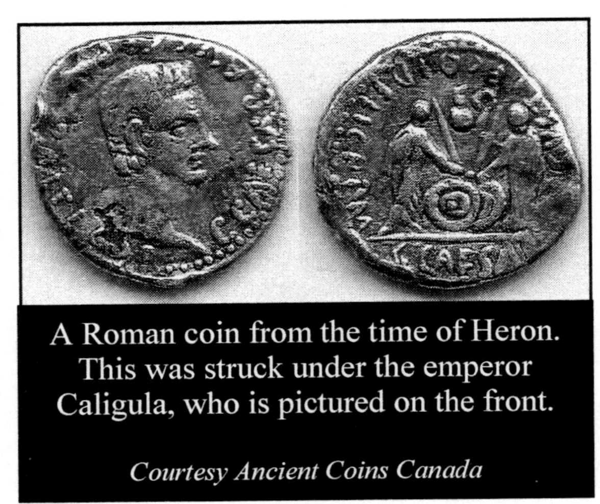

A Roman coin from the time of Heron. This was struck under the emperor Caligula, who is pictured on the front.

Courtesy Ancient Coins Canada

A statue found in Herculaneum. This shows the way people in the later Roman Empire who chose to wear togas wore them. Notice the crooked left arm. This was required at all times to prevent the toga from falling open.

Public domain photo

In addition to all these inventions, he also developed the theodolite, a surveying instrument used to measure horizontal and vertical angles. A water fountain he developed, known as Hero's fountain, used air pressure, in a pneumatic system, to create a vertical column of water which was continuously sustained.

Heron found he could make money by selling mechanical devices to churches around the ancient world. One such device was the earliest known automatic door. Near the entranceway to at least one temple stood a large metal colander and the temple official leading the guests into the building would build a fire within the metal bowl. This fire would heat water underneath the steps, which would form steam. This steam would then go into a cylinder, which, through water and air pressure, drove various gears and cables, opening the doors. As the steam cooled back into water, a counterweight pulled the doors closed once again.

Once inside many temples, it was required that a visitor wash their hands. This gave the churches and temples another way to impress believers and non alike (and raise additional revenue) with a remarkable device. This was the world's first vending machine, also invented by Heron, 1800 years before the invention of the modern vending machine. This device would dispense water that had been blessed by the local holy figures when a coin was inserted.

There was also another way that Heron invented for the temples to raise money. This was the first known mechanical con game, known as an omen machine. A small metal bird, similar to the one he used in his classes, was attached to the top of a box and the visitor would ask a yes or no question. When a coin was inserted, the bird would either sing, or not sing. Of course, the device ran on levers and had water and air pressure operate a whistle, to mimic the singing of a bird. The heads of the local temples could set the bird to sing, or not to sing, depending on which outcome to the question they desired. This device, however primitive it seems to our modern outlook, must have seemed miraculous at the time.

While in Alexandria, Heron read the works of Archimedes. These works are what inspired Heron to begin to design war machines. He designed the first primitive military robot, which would travel under its own power towards enemy troops and fire a volley of arrows into the opposing forces. He also designed a chain-driven machine gun, which fired scores of arrows harder and faster into an assembled line on the battlefield than any human could possibly hope to match. Naturally, this attracted the attention of the Romans, who began to incorporate the anachronistic technology into their forces, which provided them a further edge over anyone who dared to challenge the power of Rome.

Heron also preceded the modern age with the invention of the first automated stage scenery and even a complete play, performed only by mechanical actors in front of a vast audience. This is where his invention of computer programming really took root. Similar to an automated puppet show, the device used gears and levers to raise and lower scenery and wooden characters to tell the story of *Nautilus*. The play even came complete with sound effects, including a series of metal balls, which fell upon various plates and drums to create the sound of thunder during a storm scene. The effect on the crowd seeing the world's first fully automated, mechanized play must have been astounding.

At the same time that the Heron was designing machines thousands of years ahead of his time, work began (and was completed) on the famous Coliseum of Rome. This magnificent structure was used until the end of the Roman Empire, when it was abandoned. The name Coliseum is believed to be derived from its position, near the giant statue of Colossus, which had been constructed in Rome under the reign of Nero, who ruled from 54 to 68 CE.

The structure stood 160 feet high and was built as a four-level stadium, with seventy-six entrances on the ground floor, where people would enter and sit upon either wooden seats in the upper levels (if you were a commoner and/or woman), or upon marble located in the lower levels, if you happened to be a nobleman. Additionally, four other entrances were located in the structure, two for the gladiators themselves and two just for the emperor.

Located under the wooden floor of the Coliseum were thirty-two pens for the animals, which were raised to the arena floor by 256 choreographed Roman workers, pulling a total of twenty-eight winches. The underground of the arena also held cages for the slaves, doomed to fight (or just die helplessly) that day. This was also the area where the gladiators, trained to fight and die for Rome, spent their time before the big event. There were perhaps 2,000 such professional gladiators at one time for the Coliseum.

Over 50,000 spectators at a time could view the events happening here which included slaves fighting each other (a slave who lived long enough could win his freedom), gladiators in noble combat and starved, crazed animals mauling other animals, as well as people. Over the course of the life of the stadium, over 700,000 people died for the amusement of the spectators. The number of animals killed was even more horrific – on many days, over 5,000 animals would die in a single performance. Many of these species, brought in from such exotic locales as Africa and India, became extinct due to the blood thirst of the Romans.

The year after the death of Heron, the western Italian city of Pompeii suffered a massive earthquake, destroying much of the city. This was a wealthy city, a trading center for the Mediterranean and her soil was rich, growing a diverse variety of plants. Little did they know that the mineral-rich soil, which was her blessing, was due to the same factor that would spell the downfall of the city. For Mt. Vesuvius, which was just east of the city, was a volcano and on 24 August, 79 CE, it erupted on a warm summer day.

The morning had started like any other, as the people of Pompeii and the Roman resort town of Herculaneum nearby went about their normal routines. It was, strangely enough, the day of the celebration of Vulcanalia, dedicated to Vulcan, the Roman God of Fire. Earlier that month, all the wells in the city suddenly went dry and the townspeople, amazed, were unable to explain the phenomenon. This event should have been a warning to the people in the city of 20,000 inhabitants.

The eruption began before noontime that day and would continue well into the next day. Earthquakes rocked the city and a large pine tree shaped bellow of smoke, perhaps 30 km (20 miles) high rose from the mountain. The sea rushed away from the city in a tsunami and crashed back onto shore. Pompeii and Herculaneum were engulfed in a black, acrid smoke which blocked the Sun from the doomed inhabitants below. Hot cinders and burning rock began to crush the city, collapsing homes and businesses. Nearly two meters (seven feet) of pumice fell on the city and cooled, forming a layer of stone, upon which people attempted to walk.

At dawn the next morning, poisonous gas erupted from the volcano, choking the survivors who had made it through the day before. Over six meters (twenty feet) of ash fell that day, completely covering the city. People were shouting, trying to find their parents, children or significant others by the sound of their voice through the blinding ash which made it impossible to see.

Inside the city of Pompeii were two historians: Pliny the Younger and the Elder. As people rushed around, first uncertain what was happening and later unsure of what

Pliny the Elder

Public domain photo

Pliny the Younger

Public domain photo

to do, many headed to the beach while others ran for shelter. The wealthy huddled together with the slaves and a lucky few took to the sea. Many people, including Pliny the Elder, risked their own lives attempting to save the lives of others. In fact, Pliny the Elder lost his life, choked to death on carbon dioxide, sailing a ship to bring others to safety.

Pliny the Younger wrote two letters afterwards to the Roman historian Tacitus, in which he described the terrible destruction that he had witnessed first hand in the most famous volcanic explosion in world history:

"A black and terrible cloud, rent by snaking bursts of fire, gaped open in huge flashes of flames; it was like lightning, but far more extensive... Soon afterwards, the cloud lowered towards the earth and covered the sea...Then my mother began to beg.. me to try to escape as best I could... Ashes were already falling, but not yet thickly. .. When night fell, not one such as when there is no moon or the sky is cloudy, but a night like being in a closed place with the lights out. One could hear the wailing of women, the crying of children, the shouting of men; they called each other, some their parents, others their children, still others their mates, trying to recognize each other by their voices. Some lamented their own fate, others the fate of their loved ones. There were even those who out of their fear of death prayed for death... "[1]

[1] Pliny the Younger in a letter to Tacitus. Quoted in "The Eruption of Vesuvius", Surintendance of Pompeii.

Ptolemy c. 85 – 165 CE

Ptolemy was a mathematician and astronomer who worked in the halls of the Great Library of Alexandria during the second century CE. He also taught in the subjects of mathematics, geography and optics. His Geocentric (Earth-centered) model of the universe ruled scientific thought on planetary motion until the seventeenth century.

Ptolemy of Alexandria, from a 16th century woodcut by Theodore de Bry.

Public domain photo

Ptolemy's full name, Claudius Ptolemaeus, was likely chosen to show that he was a Roman citizen (Claudius being a Roman name), who was also a resident of Egypt (hence, Ptolemy). His greatest known work, the "Almagest", laid forth his geocentric model and accounted for retrograde cycles* with a complex system of epicycles.

In his masterwork, Ptolemy spoke of astronomy, physics, theology and mathematics. He stated that he believed that theology could not be proven, due to the nature of its subject matter and that physics could not be proven because of his belief that it was unstable. Mathematics, however, he believed to be the only true path to solid, provable knowledge.

The prosecution of Christians in the Roman Empire began in earnest, under the emperor Nero, in the year 64 CE, about twenty years before the birth of Ptolemy. A tremendous fire was started in Rome that year and it destroyed half of the homes and buildings in the city. The cause of the fires was blamed on these early Christians, although their complicity in the fires has never been proven.

When Ptolemy was in his early twenties, the Roman emperor Trajan came to power. He was a self-made aristocrat, as were nearly all of the nobility of ancient Rome. Trajan went on to name a philosophy teacher living in Rome to the position of Consul. This teacher was a man of kind spirit and an independent mind, he had one daughter and four sons and wrote over one hundred books in his lifetime. This man's name was Plutarch and his works are among the greatest sources of information about this era that exists today. He wrote forty-six examples of "parallel lives" in a series of books of the same name. Each paired one Greek and one Roman historical figure,

---

* When looking at planets in the sky over the course of a year, planets other than Mercury and Venus will occasionally seem to go backwards and create a loop, before once again going forward. This is because the Earth revolves around the Sun faster than any planet other than Mercury and Venus. If you imagine you are in a car, overtaking another car which is in another lane in front of you, first it will appear to you that the other car is going forward before you pass it, going backwards just as you pass the car and going forward again as the other car is in your rear window.

whom Plutarch felt led parallel lives. The remainders of his works were works on ethics, entitled *Moralia*.

In 107 CE, Trajan ordered the construction of the world's first large shopping mall. Located in Rome, the market, constructed nearly two thousand years ago and still standing today, known as Trajan's Forum, contained 150 merchants, spread out over five floors in one-room stores known as *tabernae*. The building was constructed of concrete and rubble between two faces of brick, which gave the appearance that the whole building was made from brick; a statement of the wealth and power of Rome, even if, as in this case, it was, quite literally, a façade. .

A Roman coin, known as a Sestertius, from 112 CE showing Trajan on the front and Trajan's Forum on the back.

Public domain photo

Concrete was a fairly new discovery at this time; and, in this, Rome had a "concrete" advantage; a local volcanic ash, called pozzolan, which dried even under water, thereby making it an especially good candidate for use in baths and aqueducts, along with helping in the construction of tremendous coliseums.

The Romans always loved a good show, or at least a good horse race. Trajan also ordered the reconstruction of what would be Rome's largest race track and arena: the Circus Maximus. Fully one third of a mile long and one and a half times as wide as a modern football field is long, the Circus Maximus held over 320,000 people at a time, twice as many as today's largest stadium.

It was built in the shape of a U and had a center barrier (known as the *spina*), which ran down the middle and stables capping off the open end. On either end of the spina were seven statues enclosed within temples. One of these temples had bronze statues of dolphins, the other of large stone eggs. As each lap of the horse race was completed, one of these statues would be removed to keep a running record of how many laps had transpired. Right at the finish line of the arena sat the emperor's box. There were other boxes, including the world's earliest known skyboxes, reserved for VIPs.

The outfit of a charioteer at the Circus Maximus. The hook on his shirt was designed to be used to cut the reigns in case of an emergency.

Public domain photo

Earlier attempts had been made building such an arena, beginning 600 years before the time of Trajan. However, these earlier versions of the Circus Maximus were constructed of wood and several times previously, the structure burned down, including in the years 36 and 64 CE. At least two other times, the stands collapsed, resulting in the deaths of many attendees.

Admission was free for the races here – a form of taxpayer supported racing, paid for by the one million inhabitants of Rome and her many colonies. No expense was spared, with the raceway adorned with statues, obelisks, columns and fountains. The largest of these obelisks was the obelisk of Tuthmosis II, taken from Thebes, at 32.5 meters (about 100 feet) high.

Racing was not the only event happening here, however. The Roman's thirst for blood sports extended beyond the confines of the Coliseum and into the Circus Maximus itself, beginning back in the time of Caesar. Large ditches, ten feet wide and 10 feet deep, were built, with rollers made of marble, at the edge of the field. Any animals on the field, ready to charge at the crowd would fall into the ditch, be unable to grab onto the free-moving marble roller and fall harmlessly (at least to the crowd, if not the animal itself) into the ditch.

Around the year 120 CE, the Roman Empire was at its largest and most grandiose. Stretching from the Atlantic Ocean in the west, to the Caucasus in the east and from Britain in the north to the Sahara and modern day Iraq in the south, it encompassed about half the land area of the modern-day United States. The area covered by this vast dominion is, today, thirty-six different countries.

An onyx cameo of the emperor Hadrian who, along with Trajan, is considered to be one of Rome's "Five good Emperors"

Public domain photo

After the death of Trajan in 117 CE, his ward Hadrian was proclaimed emperor by the army of Rome. Hadrian had been born in the area known today as Spain, making him the first Emperor born outside of the Italian peninsula – years later, he would also become the first of the Emperors to die outside of the mainland.

The Empire was plagued by small revolts on the borders of their vast dominion. Hadrian ordered the end of the expansion of the Roman Empire and withdrew the borders of Rome to the limits established during the reign of Augustus. He also built a defensive wall in the British Islands, nearly 120 km (74 miles) long, designed to hold back the tribes that were still attacking Rome in the north, including the Picts from northern Ireland and Scotland. This wall, protected by watchtowers, fortifications and traps, became known as Hadrian's Wall and parts of it still stand to this day. To further ease tensions in the outlaying areas of the Empire, Hadrian also toured his vast land, settling disputes and resolving conflicts.

The Roman control of what is now Britain had an effect in the modern day to which we give little thought. The Romans military counted a single pace as two steps – from the time one foot left the ground, until the other foot touched back down. They marked off each 1,000 of these paces with milestones (from the Latin word *mille*, for one thousand). When the Anglo-Saxons later invaded the land in the sixth and seventh centuries, they used single steps, known as yards, to pace their troops. So, they would have had 2,000 yards per mile, but they were taller than the Romans had been and it was determined that there were 1,760 yards per mile. With three feet per yard, the distance between the marker stones was designated as having 1760x3, or 5,280 feet per mile.

Convinced of their abilities and determined to build the biggest and best, the Romans set to construct one of history's most famous structures as Hadrian ordered the construction of the Pantheon (the original name of which is unknown). This temple, dedicated to all the Gods (if only to play it safe), stretched over forty meters (130 feet) across its dome and the entrance was a twin set of bronze doors, each seven meters (21 feet) high, which opened into a large, open room, under the massive dome. It had a single source of light, a round window known as an ocular, right in the middle of the ceiling. This both gave a grand appearance and allowed the dome to rest most of the stresses at the center of the dome onto a ring, encircling the ocular, for greater stability. Although no one knows who actually designed the Pantheon, the most likely architect would have been Apollodorus of Damascus, who had worked with Trajan throughout his reign and whom Trajan would later have executed.

Hadrian had been fascinated by architecture his entire life and considered himself an amateur architect. He was born on 24 January 76 CE, at a time when the Empire had a population of sixty million people. He was somber, angered easily and found competition in intellectual games. Whereas his predecessors had attempted to secure Roman security through conquest and border clashes, Hadrian tried, at least early in his reign, to find mutually beneficial relationships with the lands on his border. His reputation suffered among his subjects by the end of his reign, with some Romans believing that he had become a tyrant. His persecution of a Jewish rebellion was so intense and fought with such abandon that many believed that the entire Jewish race had been wiped out. He created a villa for himself, which he designed in the shape of an Egyptian temple, in Tivoli, a city east-northeast of Rome that was known for its waterfalls.

Most of the second century CE was a period of relative peace in the Roman Empire. Since 96 CE, Rome had been ruled by emperors who were not as nearly war-hungry as others before and after them. First, there was Nerva and then Trajan, followed by Hadrian, Antoninus Pius and Marcus Aurelius. During this time, until 180 CE, Rome was under the rule of these "five good emperors." As the time came when each one knew that a successor would need to be found, they named the person to take their place. This prevented massive power struggles from occurring and further damaging the Empire.

Ptolemy is not known to have traveled anywhere but Alexandria during his life and it is from this city that he did his greatest and most extensive work in astronomy from 127 to 141 CE. When he wrote these observations and theories down, the work was entitled (translated from Ancient Greek) "The Mathematical Compilation." Soon, it became known as "The Greatest Compilation." Later, while the Arabs were preserving the works of Ancient Greece and Alexandria, as Europe suffered through the Dark Ages, the Arabic title became *Al Megiste*, or "The Greatest." It is from the word *al-Majisti* that we get the modern title, "The Almagest." This book would be the definitive work on astronomy until the mid seventeenth century, a century after Copernicus published his Heliocentric (Sun-centered) theory in *De Revolutionibus.*

Ptolemy believed that the study of astronomy made a person a better human being, writing that it made astronomers more appreciative of natural beauty. Building on the work of Hipparchus and Aristotle, he cataloged over 1,000 stars and forty-eight constellations in the *Almagest* and carefully measured the lengths of the seasons. However, he made several errors, many of which were based on his incorrect notion of the length of the tropical year. Also, since his star catalog showed the stars in nearly the same position as those recorded by Hipparchus over 250 years earlier, he believed, mistakenly, that this was proof that the "fixed stars" do not move. We of course know today that the stars are moving at tremendous speeds, but the distance from Earth to even the nearest star is so great that the motion of the stars is nearly undetectable by the naked eye over the course of such a short astronomical time.

He also argued for a geo-centric (Earth-centered) universe, but he did so logically – if he had even an idea of how great the distances to the stars really are, his argument would have fallen apart. He tried to explain certain motions of the Moon through the further use of even more epicycles and deferents. His theory required that there was a difference in the alignments of the epicycles of Mercury and Venus when compared with the other planets. Ptolemy could only assume a divine will on the part of planets to explain their movements. He was attempting to create a model of the universe that was first mathematically viable and secondarily physically correct, but he could not get away from the ideas of both a geo-centric universe and perfectly circular orbits. Explaining the movements of the heavens with a geo-centric universe surrounded by spheres was becoming more and more difficult.

Ptolemy also wrote several other books, including "Optics," a treatise on what was known at the time about the science of optics, including refraction (the bending of light as it travels through a medium like glass, water or air), reflection and mirrors (made at the time of polished metal, not glass) and colors. Many of these ideas had been known before his time, but he set out to prove these ideas through experiment and the use of mathematics. He tested and retested as he did his experiments, for instance trying different densities of water during his experiments concerning the optical qualities of images in water. Despite his mistaken belief (which was held by most people of the time) that light rays emanated from the eye, he would lay down proofs of the basic laws of optics still recognized today in nearly identical form.

He also produced "Geography," which mapped out the locations of various well-known areas in the ancient world, using a system of latitude and longitude. For the lines of longitude, he started the numbering of zero degrees at the westernmost point known to the ancient world at his time, the Canary Islands and continued 180 degrees to China. Unlike modern cartographers, he used the length of the longest day (twelve hours at the equator, twenty-four at the poles) to measure latitude. At the time of the publication of "Geography," he had little knowledge of the true locations of most areas that were outside of the Roman Empire. Because of this, the location of most areas outside the control of Rome and even some areas well within the Empire, were badly misrepresented. He was also aware that the "global map" he produced only covered about one-quarter of the total size of the Earth. Ptolemy's book "Geography" was lost for nearly 1200 years, until it was rediscovered and translated into Latin by a Byzantine monk from Bithynia named Maximus Planudes around the year 1300. Ptolemy also expanded the earlier work of mathematicians such as Hipparchus on early trigonometry and wrote of musical theory in "Harmonica," expounding on the far earlier research of Pythagoras.

Ptolemy also wrote on astrology in "Tetrabiblos." Although he firmly believed that effects, like the tides, of heavenly bodies on the Earth were proof of more subtle influences, he never attempted precise predictions. He wrote instead of what types of effects various stars would have on the world and other general prognosticating.

Within just a few years of the death of Ptolemy, plague ravished the area of Rome. This plague was so severe and caused so many casualties that people began to wonder if the end of the world had arrived. What the Mediterranean world needed was a physician, with a brilliant mind centuries ahead of his time. It turns out this is exactly what they had, in Rome, but he was working to care for the family of the Emperor. This brings us to our next great scientist, Galen of Pergamum.

Galen of Pergamum c. 130-200 CE

Galen of Pergamum

Photo believed to be public domain

Other than Hippocrates, Galen of Pergamum was the greatest doctor and medical researcher of the ancient world. He was also the last person we know of to practice actual medical research in Europe until the Renaissance.

Human dissection, which was hard enough to perform in the days of Hippocrates, became nearly impossible in the second century CE due to ethical concerns. The Roman society, which easily condoned the slaughter of people by the thousands in the Roman coliseums would have little to no part of the study of cadavers. Because of this, little actual research on human anatomy and physiology was possible. Instead, doctors and medical researchers began to just copy each other's works, with errors compiling at each retelling. Accordingly, except for Galen, medicine began to deteriorate in this century. However, Galen was a man ahead of his time.

He even managed to perform both brain and eye surgery, which are complicated tasks even today. Galen performed a few dissections on humans, often when graves would be washed out by rains and wind, exposing a decomposing body to study. At least once, he was given a German prisoner on which to perform a live vivisection in the style of Herophilus and Erasistratus. Galen considered this to be the most fruitful route to knowledge about human anatomy and physiology, advising his contemporaries to travel to Alexandria to see the insides of human bodies. He developed special instruments for surgery and his study of bodies, many of which remain essentially unchanged in the modern day. However, the opposition to these practices that had begun under Herophilus and Erasistratus became even stronger in the days of Galen. His contemporary, the Christian writer Tertullian, in "On the Soul," would describe Galen of Pergamum as "that doctor or butcher who cut up innumerable corpses in order to investigate nature and who hated mankind for the sake of knowledge."[*]

Born in the city of Pergamum, which is current day Bergama, Turkey, sometime around the year 130 CE, he was the son of Nicon, who was an architect. Nicon schooled his son in mathematics, logic, grammar and philosophy. After the death of Nicon in 148 or 149 BCE, Galen went to study in Smyrna and Corinth and later he left for Alexandria. It was here that he would do his greatest research and afterwards, in Rome, he would achieve his greatest level of fame. Galen spoke Greek

1 Quoted in G.E.R. Lloyd, "Greek Science After Aristotle." 1973.

most of the time, which at the time was considered a more advanced language for people of medicine than Latin.

Greece had always had great respect for good doctors, but Rome was a different world. The upper class Romans retained good health care, largely, while the lower classes were treated by slaves, freed slaves and foreigners who had not been trained in Rome or Alexandria. Even Galen himself considered medicine at the time as corrupt,

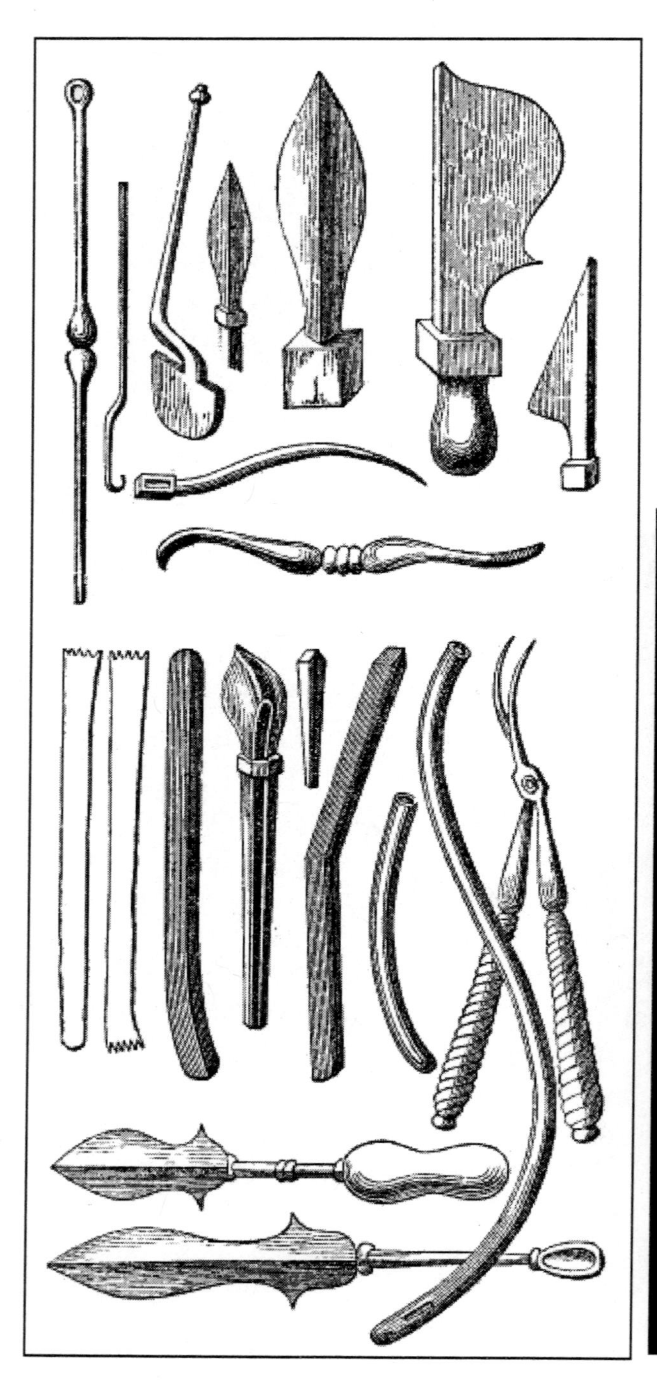

First century CE surgical tools, of the type likely used by Galen.

TOP ROW:
(Vertical) 6" probe, 4 ½" probe, cantery, lancets (2), copper knives (2)
(Horizontal) 3" iron needle, elevator (for raising the area around surgery)

BOTTOM ROW:
(Vertical) Forceps (three styles, the first is in two pieces), a tool used for removing hairs by the root, a bent tool used for extracting foreign bodies from a wound or the throat, 4" female catheter, 9" male catheter, a dental tool of uncertain use
(Horizontal) Spatulae (2)

Modern surgeons today would be able to recognize the use of several of these tools.

Public domain photo

as he worked to convince other doctors to try to avoid gathering wealth through their profession. Thus, the people of Rome began to rebel against the medical establishment and most considered the profession to be of very low status. One woman, Julia Balbilla wrote, in July 138 CE:

"They travel for three days from Rome to the coast in search of a cure, their

desire for health equaled, if not frustrated, by the physicians yearning for weath."[1]

Galen returned to Pergamum in 157 CE and took up residence in a gladiator school there. This experience in real-life injury and wounds would provide Galen with the experience he needed to learn more about the human body and to try different methods of treating serious wounds. He regarded these injuries as "windows to the body."

In the year 162 CE, he moved to Rome, where he worked as the doctor to the emperor Marcus Aurelius, who had just ascended to the throne the year before. Just a few years after he got there, Rome suffered through a devastating plague that had many people of the time wondering whether the end of the world was at hand. While the people of Rome agonized through the deaths of family and friends, watching ten percent of the population of the city succumb to the illness, Galen was treating the emperor and his family.

While he was in Rome, unable (largely) to obtain human cadavers for study despite the plague, he performed vivisections on Barbary apes. He considered these to be the animal closest to humans and he had them drowned for the purpose. He also experimented on dogs and other animals, attempting to ascertain the workings of the human body from the study of non-human mammals. This would lead to several errors in thinking, as he mistakenly believed the human liver has five lobes like a dog and that the human heart has two chambers (only fish have a two-chambered heart) and not four (such is found in all mammals and birds). Heron also disagreed with the idea, proposed by Erasistratus, that muscles alone were responsible for the movement of food throughout the digestive system. Although his dissections showed that such movement occurred, he also believed that the food had an attraction toward the stomach, which he called *Holke.*

His greatest contribution to medical literature was his seventeen-volume treatise "On the Usefulness of the Parts of the Human Body." Working from the theories of Hippocrates, Galen believed that the body exhibited four humors, which he associated with each of the four ancient elements (earth, air, fire and water). Galen believed that the liver and veins create blood, theorizing further that the arterial and venous systems to carry blood were separate systems, independent of one another. Because of this, he did not recognize that blood circulation occurs within the body and he was opposed to the practice of using tourniquets to stop massive bleeding. In fact, Galen was a proponent of bleeding as a method of draining foul humors from the body of ailing patients. His idea of bleeding as a path to wellness would remain entrenched within the medical community until the Renaissance.

Despite these errors, Galen made tremendous strides toward understanding the nature of human anatomy and physiology, including identifying the brain as the organ that actually controls the mind. Galen divided the different parts of anatomy into

---

[1] Elizabeth Speller. "Following Hadrian." 2003.

homogenous parts, such as blood, from functional parts such as the hand and foot. He was also the first to determine that the arteries carried blood and not air, as had been believed and he did extensive work with nerves (discovering the so-called Galen's nerve), as well as the heart and brain. He became so successful at treating patients that people began to believe that he was practicing magic. Even his practice of taking a pulse, known since the time of Herophilus, was considered divination by some outsiders.

Galen was also a prolific author, having a goal of writing at least three pages a day during his life. Many of these works were destroyed in a fire at The Temple of Peace in 191 CE. Although what remained of this remarkable body of work was the greatest body of medical knowledge in the world to that date, it also spelled the downfall of medical research until the Renaissance. For Galen was also a studious self-promoter, working to increase both the popularity of his works and with that, his own fame. The books that Galen had produced were the largest, most complete body of medical research in that day and this led other doctors to quote Galen, rather than to produce their own research. Medical knowledge became disseminated through the production of pamphlets quoting Galen, rather than original material. With each re-telling of the words of Galen errors compounded and what doctors were learning drifted further from the truth. No doctor in the Mediterranean felt that they could compete with the great Galen and so medical knowledge tapered away until it had devolved into little more than mysticism and superstition.

Little is know about the death of Galen, including the year in which he died. He passed away sometime early in the third century CE, possibly as late as 216 CE, having lived to his mid-eighties. With the death of Galen and the coming fall of the Roman Empire, ancient science would itself become ill and fade away. But, like a dying star, the death throes of ancient science were accompanied by moments of its brightest light.

The Smothering of the Flame

The library and its contents were to burn several times throughout its history. The first of these times was in the year 48 BCE, when Caesar (supporting Cleopatra VII against her brother, Ptolemy XIII) possibly set his fleet on fire to prevent the ships from falling into the hands of the Egyptians. This fire then may have spread to the Alexandrian ports, destroying up to 400,000 scrolls (more likely 40,000) that had been placed on the docks for shipment to Rome.

Seven years later, Anthony gave a gift of perhaps 200,000 scrolls to Cleopatra in an attempt at recompense this loss, but the library would never again regain its former glory. At this time, the Roman Empire is thought to have had perhaps fifty or sixty million people; this is roughly the population of Ethiopia, the United Kingdom or Iran today.

At the beginning of the third century CE, Rome came under the control of the tyrant Caracalla. Caracalla was a nickname given to him due to his habit of wearing a long tunic, popular in Gaul, known as a caracalla (his real name was Marcus Aurelius Antoninus). Born in Gaul (in the city which is today Lyon, France) in 188 CE, Caracalla took the throne with his twenty-two year old brother Publius Septimius Geta upon the death of their father in 211 CE. Within one year, Caracalla was responsible for the death of his brother and thousands of Geta's followers.

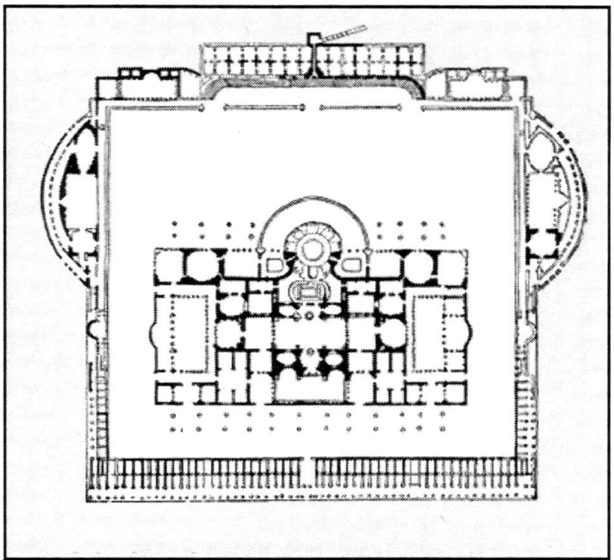

He crushed dissent in Alexandria by initiating a bloodbath in the city; the city which once flourished from free discussion and reasonable discourse became a place where people were afraid to speak their minds, terrified of the powerful force of the Roman Empire under Caracalla.

He also spent tax money like a drunken sailor on leave. Highly unpopular and known within and without the Roman Empire as a cruel, barbaric leader, Caracalla undertook massive construction projects, believing this would endear his people to him. Instead, it had the opposite effect.

Despite the fact that Rome had over 1,000 public baths within the city limits themselves, Caracalla began the construction of the greatest bathhouse of all, larger than anyone had ever seen before -

A scale drawing of the bathhouse of Caracalla. The cost of building this one bathhouse was greater than all the other expenses of Rome combined.

Public domain photo

an überbadhaus, if you will. This copious construction, built in 212 CE, was completed by over 16,000 workers, laboring all hours of the day and in all weather conditions, without reprieve. It enclosed more space than Saint Peter's Basilica (also in Rome) would when that structure was constructed in the sixteenth and early seventeenth centuries CE and the land it was on stretched for twenty acres. Holding over 2,000 people at a time, the baths were warmed by fifty fireplaces, whose heated air was carried by terra cotta tubes that were built within the walls of structure itself.

The project was so grandiose, that at its height, the construction of this one bathhouse cost as much as *every single other expense of the Roman Empire combined.* This wound up causing massive inflation and unemployment within Rome, raising poverty and decreasing what little support Caracalla retained in the Empire.

In 217 CE, Caracalla was assassinated in Mesopotamia by his own bodyguard, the fifty-three year old Marcus Opelius Macrinus, who would then go on to succeed him as emperor. The Alexandrian Museum (and the Royal Quarter) would be partly destroyed over five decades later during the power struggles that continued to shake the Roman Empire.

By early in the third century CE, the Roman road system was at its peak of construction. The roads were expanding at the rate of one mile every three days. The roads were all straight due to the fact that the surveying instruments of the day could not see around corners. Thus, when they needed to build through a mountain pass, or rough terrain, they would design the roads in such a way that they zigzagged around the obstacles or difficulties, or went right through them. These roads allowed the Romans a tremendous advantage when moving troops. The Roman legions could travel along this road system at twenty miles a day, provided they spent seven hours a day moving. Ordinary people, not encumbered by arms and military supplies, could speed along these roads at up to one hundred miles a day.

Around 250 CE, Alexandria saw another great mathematician – Diophantus of Alexandria. He wrote a thirteen volume set of books, entitled Arithmetica and became known as the father of algebra. In Arithmetica, Diophantus both united early works of previous mathematicians and expanded greatly upon the earlier works. He was also the first person to introduce special symbols for unknowns (such as X or N) into equations that we still use today in algebra. Today, only six of the original thirteen volumes is still extant.

Another devastation befell the Great Library in 270 CE when Zenobia, the queen of Palmyra (part of what is currently Syria) who had named herself the rightful heir of Cleopatra captured all of Egypt and parts of Asia Minor in 269 CE. Soon, she also declared her lands independent of Rome and went to war against the emperor Aurelian. During the siege laid by the forces of Aurelian upon the army of Zenobia, much of the main library was destroyed. After this event, the newer daughter library became the main building for the library complex.

The Roman Empire split for the first time in 285 CE, under the emperor

Diocletian, who took control of the eastern half of the formerly great empire. In 303, he would reinstate the persecution of Christians in an effort to unite the Empire under Pagan theology.

In 307 CE, Constantine became emperor of the Western Empire. He had been born in Nis, which is currently part of Serbia, in 274 CE. The son of The Western emperor Constantius I, he traveled to Britain in 306 CE in order to assist his father with a battle there. His father died that year and the soldiers he had fought with there appointed Constantine emperor. However, there were others who where vying for the throne and the Western Empire had several emperors during this time. The first year, 306, saw three emperors, Constantine, Severus II (for one year total) and Maxentius (who lasted until 312). The following year Maximian (who had been emperor before Constantius I) came back to power for one year. Certainly, by the year 324 the power of Constantine was assured.

One battle against Maxentuis at Milvian Bridge, near Rome, in 312 CE would have great consequences for history. Constantine believed in the idea of Henotheism, that the Sun (as the Roman God Sol) was the visible manifestation of a higher guiding power called *Summus Deus*. He even claimed to have been visited by the Sun God while in Gaul in 310 CE. Preparing to go into battle against Maxentuis, he dreamt before the conflict that Christ had appeared to him, ordering him to inscribe the first two letters of his name (which are XP in ancient Greek) upon the shields of his soldiers. The next day, he stated he had seen the sign of the cross emblazoned across the Sun with the words "*In hoc signo vinces*" ("In this sign you will be the victor"). After defeating Maxentuis that day, Constantine converted to Christianity and along with his co-ruler Licinius, issued the Edict of Milan in 313 CE. This edict ended the persecution of the Christians within the Empire and began an infusion of money from the public treasury of Rome to the Christian churches. It was this event that began to turn the Roman Empire into a Christian state.

The Roman Empire stay spilt into two parts until 337, when Constantine II defeated Licinius in battle and the empire briefly reunited. However, this would only last for twenty-seven years, until 364, when the emperor Jovian (who had only been emperor for one year) passed away and the empire once again fractured, leaving an Eastern and Western Roman Empire in its wake.

In 391 CE, the Eastern Roman emperor Theodosius decreed the banning of all Pagan works. Following this lead, the Serapeum and the daughter library were destroyed at the hands of Theophilus, who was bishop of Alexandria from 385 to 412 CE. Objects taken from a Pagan temple were paraded around the streets in mockery and ridicule. Soon, a riot broke out during which several Christians were killed. Theophilus declared that those who had been slain were martyrs, killed in the name of the one true God. The Pagans took shelter within the Serapeum and fortified it against attack. Several Christians were now held prisoner within its walls. Inside of the building was a tremendous statue of Serapis, made of several different types of metal as well as stone, that nearly filled the inner hall. At first reluctant to destroy the tremendous statue, fearing the Earth would open up and swallow them all if it were to be damaged, one member of the Christian mob finally gathered enough courage to strike at the effigy with his ax, striking it in the jaw. Sure enough, no hole in the ground opened, and soon other soldiers began to hack away at the behemoth figure. It was not long before it was destroyed, and a large group of rats came up from underneath its base. The head of the statue was paraded through the town and what remained was burnt in front of the Pagans who had worshiped this God of healing. Busts and statuettes of Serapis found throughout Alexandria were smashed and the angry mob painted crucifixes in their place.

The entire Serapeum was then destroyed, along with other temples and statues that had filled the city. Nearly every piece of metal that had once held significance for the Pagan worshippers was melted down and recast into goods for the Christian churches. Only one statue was left by Theophilus as a reminder to future generations that such Gods were once worshipped in this place. Nearly every scroll was either destroyed or taken by the rampaging mob. Contemporaries who saw the destruction left in the wake of Theophilus' devastation stated that only the floors of the Serapeum were left behind by the angry mass, and those were only left because they were too heavy to carry off. The final destruction of the library and the brutal murder of its final director Hypatia would take place in the year 415 CE[*] at the hands of Cyril, the successor to Theophilus. Despite this act and a terrible persecution of the Jewish people in Alexandria, the church made Cyril a saint.

Constant warfare, religious intolerance and a massive bureaucracy attempting to control science had stifled knowledge and learning in the ancient world. Medical researchers could not practice their craft due to the prohibition on human dissection, asking questions about the objects in the sky was considered sacrilegious and women like Aglaonike and Agnodice were considered to be impure if they dared to enter the field of science. The final destruction of the library and its accompanying scrolls were the final acts in a series of events that would plunge the world into the dark ages from which we would not awake for over one thousand years.

There was one more shining light on the horizon however, a scientist and mathematician who would buck all the trends, question those things which were not supposed to be questioned and live the life of a Pagan in a world where Christianity was the law of the land. Born in Alexandria, this scientist would be the last great gasp of air for science that was quickly being smothered. As if this were not enough, this last great scientist of the ancient world was also a woman – Hypatia of Alexandria.

---

[*] The story of the destruction of the Great Library is one of the most hotly debated topics in ancient history and there is reason to believe some part of the library may have stood until 640 CE. However, in 415, most of what was left of the library after Caesar's fire was torn down and science at the library was finished, even if a few scattered rooms and scrolls may have still been waiting to be used.

Hypatia of Alexandria 375 - 415 CE

Hypatia of Alexandria

Public domain photo

The last of the directors of the Great Library, Hypatia, was a woman ahead of her time. For that, she paid dearly. She was an astronomer, physicist, mathematician, philosopher, author and political activist. She was also a Pagan, in a world ruled by the increasingly Christian Roman Empires.

Hypatia was born in the year 375 CE* and her father, Theon, was director of what remained of the Great Library. It was with her father that she helped to compile older mathematical works and a more popular edition of Euclid's "Elements," which was used almost exclusively by Greek teachers after her time. Together, they also compiled a commentary, in eleven parts, on Ptolemy's "Almagest."

Theon was mainly a mathematician, but Hypatia had a more inclusive view of science and philosophy, studying fields far beyond Theon's scope of study. She would walk throughout the streets of Alexandria, regaling people with her interpretations of early philosophers, including Aristotle and Plato. Wherever she could find an audience ready to listen, she would speak to them of her latest theories and discoveries, in her own characteristic style. She was well-spoken and articulate, and she was not afraid to speak her mind, even in the presence of knowledgeable men. This behavior was almost unheard of, even at this late date, and it was this independence and eloquence that made many people in Alexandria begin to look on her with great reverence.

One work normally attributed to her was entitled "The Astronomical Canon." Although none of her original works remains today, she is believed to have also done work on conic sections (in a commentary on "The Conics" by Apollonius) in which formulas for ellipses (including circles), parabolas and hyperbolas are derived from mentally slicing one (or in the case of the hyperbola, two) cone(s) in various ways.

While in her mid-twenties, she was named head of the Platonist school of philosophy, where she taught Neoplatonism. This was a branch of philosophy, developed in the third century by Plotinus of Egypt and Iamblichus of Syria. This school, which taught the philosophy of Plato, also mixed in ideas from the Stoics and

* There are conflicting reports as to when Hypatia was actually born and how old she was when she died. However, the idea that Hypatia was older than the dates I use (some people believe she was born 20 years earlier) are mostly based on the supposition that her students *had to be* younger than she was, as their instructor. I find this a false presupposition, as I have had several students older than I and without further evidence to the contrary, I am going with the dates in earlier reports.

Aristotle.

When Hypatia was thirty-five, for the first time since its founding, the city of Rome fell to an outside conqueror. Through all the conquerors that attempted to defeat Rome, no external enemy had managed to take Rome itself. That was, until 410 CE, five years before the brutal death of Hypatia, when Rome proper was sacked by the leader of the Goths, Alaric I.

The Goths ruled the area that is now the Ukraine and Alaric was born there, on the lower Danube River, in 370 CE. The year he was born, the Goths divided into two factions – The Eastern Goths, or *Ostrogoths* and the Western Goths, otherwise known as the *Visigoths*.

During the decades leading up to the birth of Alaric and in the boy's youth, the Huns were pressing westward, thereby pushing the Visigoths further west, into the hearts of the two Roman Empires. During that time, the Visigoths played a small role in the Roman military as support troops to the regular army.

In the year 395 CE, the Visigoths named Alaric their king and broke their allegiance to Constantinople. He marched his troops into Greece and conquered Argos, Corinth and Sparta. Athens only avoided falling to Alaric's forces by surrendering a large ransom to the Visigoths.

The Western Roman general Flavius Stilicho eventually defeated Alaric and he retreated to Illyricum (modern-day Albania), where he ruled as prefect under the leadership of the Eastern Emperor Arcadius.

Seven years after being given the throne, in 402 CE, Alaric invaded Italy and was again defeated by Stilicho. After time, Alaric agreed to join forces with the Western Roman Empire (now under the rule of Honorius), in a planned attack against the Eastern Roman Empire.

However, in 408 CE, the Eastern Roman Emperor Arcadius passed away and so Rome abandoned her plans to invade the Eastern Empire. As compensation, Alaric demanded two tons of gold ($427.2 million dollars at today's prices) for what he perceived to be a breach of contract. Alaric's nemesis Stilicho convinced Honorius to

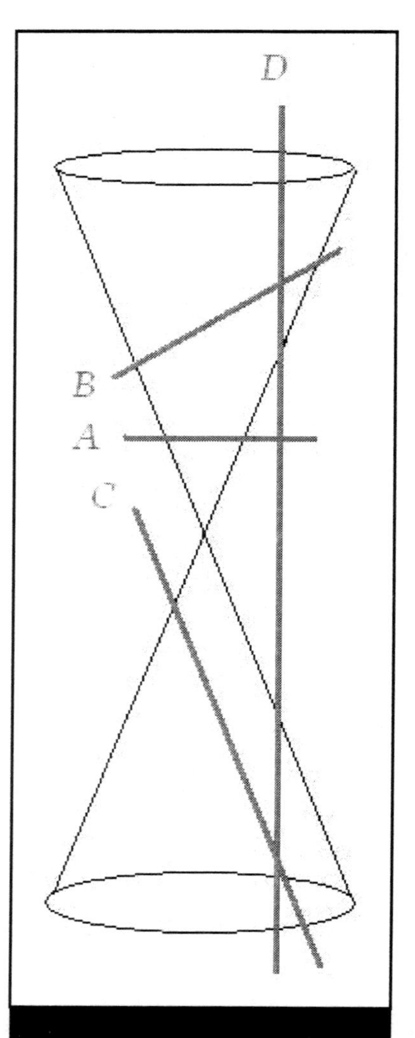

Conic sections:

Two cones put tip to tip could be sliced four different ways, producing different shaped cuts:

A) Circle
B) Ellipse
C) Parabola
D) Hyperbola

NOTE: To get a parabola, the slice is made parallel to the side of the cone as seen from the side.

agree to the demand, but soon Honorius had Stilicho executed and Rome reneged on the agreement.

This caused Alaric to once again invade Italy and the Visigoths began a siege of Rome. The Romans paid a hefty tribute to Alaric, but the peace that had been paid for would not last long. In the year 410, Alaric became the first and only outside army to conquer Rome itself. He did this after a long siege of the city, which turned into a stalemate. Alaric knew he could not take the city by force without resorting to out-of-the-box thinking. What Alaric did was to essentially use a Trojan Horse, without the horse. He gave notice to the Romans that he was surrendering, leaving three hundred of his best fighters as payment to the Romans in compensation for the war. The Romans happily took the fighters inside of the city gates to be given to members of the Senate as spoils of war.

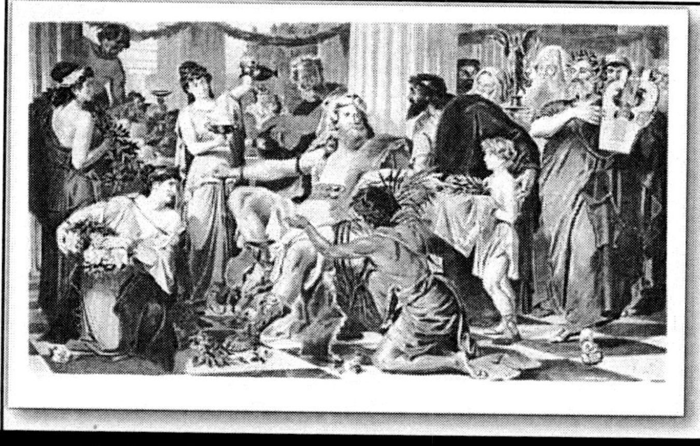

Alaric I, in an 1894 photogravure of an oil painting by Ludwig Thiersch.

Public domain photo

Sure enough, there was a plan. The army of Alaric, which the Romans had thought left the area was, in reality, barely out of sight, waiting for the time to pounce. At a pre-determined time and date, the soldiers inside the city met after lunch while their new masters were taking naps, gaining control of one guard tower. Then, like the soldiers in Troy, they opened the main gates to let their main forces through the door into the city.

Suddenly, even the center of the new (and then largely inconsequential) papal sect of the Christian religion (soon to become the Catholic Church) was under the command of Alaric. Strangely, Alaric's forces only stayed in Rome for three days before leaving the city completely. One idea for this behavior is that Alaric did not truly wish to conquer the Western Roman Empire, but to instead show to the Romans that he and the rest of the Visigoths, were a powerful force to be reckoned with and respected.

Now Alaric set his sights on Sicily and Northern Africa. However, at the beginning of what would have been an enormous campaign, a tremendous storm wiped out many of Alaric's forces and he was compelled to abandon the quest for Northern Africa and Sicily. Back at home, Alaric took a new wife and they spent their wedding night drinking and making love. When she woke the next morning, she found Alaric dead beside her, his face covered in blood. It is now believed that Alaric developed a bloody nose during the night and drowned in his own blood.

However, this had an unexpected consequence, as the followers of the pope in

Rome, eager as any in the city to see the death of Alaric, proclaimed his sudden death a miracle and a work of God. This helped swell the number of adherents to the papacy and expanded the power and reach of the what would become the Catholic Church. After all, people reasoned, if God killed Alaric for the Church, then following the teachings of the pope must be the true religion. This lead directly to the kind of dogmatic adherence we would see to the Vatican during the Dark Ages that would soon fall upon the world.

Alaric was seceded by his brother (or perhaps brother-in-law), Ataulf, who rejected the idea of expanding southwards and instead marched the Visigoths westward, into Gaul (modern day France), where they took up residence.

Hypatia was now well-known throughout the city for her great beauty, intellect and charismatic style of speaking and teaching and her fame had reached near-mythic proportions. She was known to wear her hair parted down the middle, dressing in purple robes (the traditional color of royalty) with jewelry of pearls and emeralds. Her melancholy personality was accentuated by her sad, distant eyes that left people with a feeling of respectful distance from her remarkable mind. While teaching at the school, she developed friendships with people who held political power and she began to anger the church, who considered her teachings to be Pagan and therefore, dangerous. The early twentieth century English author Thomas Little Heath wrote of Hypatia:

"By her eloquence and authority ... [she] *attained such influence that Christianity considered itself threatened.."*

Yet, the belief that Hypatia's teachings were dangerous to the Christian church did not find root with one of her most famous Christian students, Synesius of Cyrene, who later became the Bishop of Ptolemais. He wrote to her with great admiration and respect, as one might expect from a loyal student to a beloved teacher. In these letters, Synesius asks Hypatia for her assistance in building a hydroscope (a device used to see deep under water) and a astrolabe (which is used for measuring the angle between a star and the horizon, to measure one's latitude).

Nevertheless, this approval and esteem from one devout Christian did not extend to the man who would spell her downfall, Cyril. One year older than Hypatia, Cyril was a theologian and a prolific author, born in Alexandria in the year 376 CE. He was a nephew of Theophilus who had earlier destroyed the Serapeum at the Great Library. He was best known in his time as being an outspoken opponent of the idea of Nestorianism (which was named after the patriarch of Constantinople, Nestorius). This was the idea that Jesus of Nazareth was two separate people: one divine, the other human, and Cyril found the idea repugnant.

In 412 CE, Cyril was elected patriarch of Rome and he immediately began a persecution of those he considered hostile to Christian beliefs. His followers raided, looted and closed churches and began a violent attack on Jewish households and

Cyril of Alexandria – The prime suspect in plotting the murder of Hypatia

Photo believed to be public domain

businesses in Alexandria - a fifth century Kristallnacht, repeated many times over. This hatred boiled over, after steeping from the angry discourse centuries before from Helicon, a slave to Caligula, who directed his wrath at the Jewish population in Alexandria. Soon, the Jewish population of Alexandria fled from the city in terror. Passing by the house of Hypatia one day, Cyril found a great crowd of people gathered around her home. Cyril asked the assembled crowd why they had gathered. When told that they had come to see Hypatia, who was about to come out to greet them, Cyril was enraged. It was immediately after this event that Cyril began to plot her murder.

However, there was another reason that Cyril had such a distaste for the Pagan philosopher. Hypatia was friends with Orestes, a prefect in Alexandria with whom Cyril had a blood feud. Christians in the city disliked Orestes and the rumor that Hypatia was responsible for the bad blood between Orestes and Cyril began to spread throughout the city. Christians in the city accused her of practicing magic and of keeping the Governor of Alexandria from attending church through a magical trance upon him by Hypatia. Moreover, she also attended and organized large public dancing festivals in the city, which was considered unchristian.

The Jewish population of Alexandria began to fight back against the Christian persecution, but this only hardened the resolve of the followers of Cyril to fight Pagan and Jewish influence in the city. It was during the course of one of the riots against the Jewish population in Alexandria during March 415 that Hypatia was traveling back to her house in a chariot.

On her way home, the legions of Cyril, lead by a man named Peter, fearing the power that Hypatia had assembled, were there to meet her. She was dragged from her chariot and brought to the public church of Caesareum, where she was stripped naked and skinned to death with sharp edged objects, perhaps oyster shells, or ceiling tiles.* Her body was then literally torn apart and her remains were brought to Cinaron, where her body and works were burned.

The Christian mob, going against every central teaching of Christianity, did their best to destroy both her body and everything related to Hypatia in an attempt to wipe her teachings from the pages of history. The Western Roman emperor, Honorius, was horribly upset by this assassination and would have prosecuted Cyril for the murder of Hypatia, but bribes given to the right people in power ended the justice that would have fallen upon Cyril.

* Ancient sources list the tools used to skin her alive as *ostrakois*, a word which meant both oyster shells and brick ceiling tiles, so it is uncertain exactly which type of object they used.

After this series of riots, culminating in the horrific murder of Hypatia, the intellectuals in Alexandria fled from the city and Alexandria would never again regain her previous glory.

Fade to Dark

By the third century CE, scientists working in Alexandria and other areas became more concerned with preserving old knowledge rather than acquiring new knowledge. Dissections, for instance, were performed in rare cases, but they were used as a method of teaching rather than research. Doctors still continued to practice their art, but added little to the knowledge that had already been found by previous researchers. Like medicine after Galen, science as a whole became disseminated mainly through the summaries and pamphlets that were all the rage at the time.

After the final destruction of the Library of Alexandria and the brutal murder of Hypatia, the Dark Ages would fall upon the world. Just eight years after the death of Hypatia, Theodosius and Honorius wrote that they believed the entire Pagan population of Alexandria had been destroyed. Rome fell in the coming decades and what knowledge remained from the Ionian awakening and the Library of Alexandria was stored by Arabs in Toledo, Spain and then brought back to the Middle East. However, the Arabs did little with this information and did not create or discover much more than what was in the remnants of the great works of the Aegean and Alexandrian philosopher/scientists.

There was, however, one notable scientist working in Alexandria after the death of Hypatia. His name was John Philoponus, and he wrote several commentaries refuting Aristotle's flawed notions of movement in his work "Against Aristotle." Philoponus was born in 490 CE and lived to about 570 CE just after the Roman Empire fell apart. Philoponus saw the different colors apparent when looking at planets and theorized that they were made of different materials, which is more or less correct. He correctly deduced that a medium such as air was not responsible for the forward movement of an object such as an arrow, as had been claimed by Aristotle, but that such a medium would only hinder its movement. He also deduced (correctly) that the difference in times between two falling objects is not in proportion to their masses as Aristotle had claimed. Although he failed to realize that objects in a vacuum would fall at the same rate, Philoponus used experiments and careful observation to show many fallacies in Aristotelian doctrine.

The Academy founded by Plato in Athens was closed in 529 CE by the Byzantine Emperor Justinian, who also forbade Pagans to teach, threatening them with crippling fines, confiscation of property and even exile if they dared to defy the order. Anyone who dared to teach without being baptized was now a criminal and subject to prosecution by the Christian Byzantine Empire that had evolved from the Western Roman Empire.

The works of the ancient Greek scientists did not lay completely dormant during their stay with the Arabs, however. Some notable exceptions to the generality that the Arabs who held on to this work did not expand it were al-Hazen, who worked in the Middle East on furthering the Alexandrian work on optics and al-Khwârizmî, who was an Arab mathematician who lived in the ninth century CE. Meanwhile, Europe stagnated through a thousand years of superstition and persecution.

It is interesting to think of what may have been if the Dark Ages had never occurred. Science would be at least a thousand years ahead of where we are now technologically. Would today's ultra-fast four Gigahertz systems appear as archaic as horizontal looms and iron plows (both eleventh century inventions)? Would, today, cancer and AIDS appear nearly as ancient and long forgotten as the bubonic plague?

On the other hand, could things have worked out for the worse? After all, with an extra thousand years of technology, the Vikings under Leif Erickson might have had nuclear weapons and we can only imagine the terrible forces of destruction that would have been available to Napoleon and Hitler. These are questions to which we will never have the answers. With the way technology and science are expanding today, it seems hard to contemplate what our society would be like today, had the Dark Ages not occurred.

Just as the creation of science was not due to a single, or even a handful, of reasons, the end of ancient science was also caused by a number of different forces that together temporarily extinguished the light of logical reasoning. No single one of these factors could have been the single cause of such a horrific loss to the world. Each of them alone would have been a setback to the library, but one from which it could have eventually recovered. Together, they spelt doom for the buildings which held the scrolls and the classrooms, but more importantly, they worked together to turn the burning flames of science into embers for a thousand years.

Many authors today are quick to blame the slave trade and the accompanying plentiful supply of labor for the failure of popular science to take hold in these early days. Certainly, this was a part of why these people did not grasp the magnificent future ahead of them if they had followed the path of technology and science. However, as we have seen, this was not the entire story.

The Roman Civil War, which spilled over into the Alexandrian War, played a tremendous part in the coming of the Dark Ages. There was no way that a giant empire such as Rome could fall and not have repercussions around the world, particularly within their sphere of influence, as Egypt was after the Alexandrian War.

A lack of knowledge about real economic forces and prices made it nearly impossible for investors (and there were plenty available) to see how devices like Heron's aeolipile could make far more money for the investors than land ever could. Therefore, instead of investing in new technologies, the usual strategy was to put your money into a box for safekeeping. Even when land was available for sale (usually a safe investment), little mention was ever made of the production capabilities of the land, or which crops might be planted to produce the greatest revenue. A potential buyer looking for land might look for how close it was to the people they knew, what sort of crops were growing there at the moment (preferably crops that the buyer liked to eat), how much sunlight the property received, etc. There were no future earnings to look at, associated costs, development costs, or anything of the sort. The result: A society full of investors, who never saw the advantage of investing in technology.

Another reason was widespread religious intolerance. Science was born in Ionia, where hundreds of religions and a lack of a nearby bureaucratic government made questioning nature acceptable and tolerated. Nevertheless, as soon as it spread, even to Athens (with a polytheist religious tradition), questioning of nature was interpreted as questioning of the highest authorities and was suppressed.

The people of Greece, from the Fourth to Third Centuries BCE, began to accept differing views on religion and they saw the greatest flourishing of science in the ancient world. Science once again reached great new heights (Heron, Galen, et al.) when Rome was divided between polytheists and Christians. As soon as Christianity became law of the land and received state endorsement and financial support, we once again saw science suppressed, culminating with the final destruction of the Library of Alexandria.

However, Christian writers at this time also had differing ideas about the study of nature, from the height of tolerance of Origen, who wrote in the third century CE:

"I would wish that you should draw from Greek philosophy too such things as are fit, as it were, to serve as general or preparatory studies for Christianity and from geometry and astronomy such things as may be useful for the interpretation of Holy Scriptures."[1]

Down to the intolerance expressed by Augustine, who wrote, around the year 400 CE:

"Nor need we be afraid lest the Christian should be rather ignorant of the force and number of the elements, the motion, order and eclipses of the heavenly bodies... It is enough for the Christian to believe that the cause of all created things... is none other than the goodness of the Creator."[1]

In the fifth century CE, the Christian church began to look on the ideas of Aristotle with some interest. After all, his idea of the inner spheres moving due to action on the outside sphere, suggests an initial mover. In the days of the final downfall of Rome, long after Aristotle, that prime mover was deemed to be the Christian God. In the Old Testament, God is said to have stopped the Sun and Moon, so that Joshua might successfully wage battle against the Amorites (from Canaan and Babylonia):

"Then spake Joshua to the LORD in the day when the LORD delivered up the Amorites before the children of Israel, and he said in the sight of Israel, Sun, stand

---

[1] G.E.R. Lloyd. "Greek Science After Aristotle." 1973.

thou still upon Gibeon; and thou, Moon, in the valley of Aj'alon. And the sun stood still, and the moon stayed, until the people had avenged themselves upon their enemies.".

King James Bible: Book of Joshua 10:12-13

According to the interlocking-gear mechanism of Aristotle, God could have stopped the Sun and Moon, easily enough, by stopping the outermost sphere. However, that would have also stopped every other sphere inside, similar to the way that every gear of an analog watch would stop if one were foolish enough to crazy glue one main gear to the side of the watch. Accordingly, by Aristotle's hypothesis, every change would have also stopped, including change on the Earth such as the movement of Joshua's troops.

This apparent contradiction when attempting to meld scripture and Aristotle occurred to Augustine of Hippo, who was about twenty-five years older than Hypatia. He had been born in Tagaste, Numidia (currently part of Algeria) on 13 November, 354 CE, the son of a Pagan father and a Christian mother. He reasoned that time was not dependent upon the movement of the objects in the heavens, and so he was able to temporarily quell the discomfort that such contradictions brought to the study of philosophy and theology. By the thirteenth century, additional contradictions were found and this time, the genie would not be put back into the bottle.

The fourth of many reasons for the coming of the Dark Ages is also the most insidious. Throughout the entire period of the Greco-Roman era, most of the scientists did little to popularize science. There was no major figure like Carl Sagan or Steven Hawking to bring science to the masses. The result of this was a public that was apathetic towards science and unwilling to support those who were questioning nature when governments found their questions uncomfortable.

Perhaps an argument could be made that some of the devices created pleased the masses, such as Heron's mechanized stage machinery, or his automatic doors, or the Antikythera mechanism. However, these devices were all created to awe the audience, not to demonstrate the power of steam or differential gears. Perhaps if mass production had been developed and millions of copies of these remarkable devices had been in the hands of a curious public, greater uses would have been found for these inventions.

Over the course of the last 1,600 years the shorelines around Alexandria and the Nile have both changed shape. Most of what was built in the ancient city, including Cleopatra's palace, now lies underwater. Parts of the Serapeum have been found, but the most of the rest of the library and its attendant museum remain lost. A team of Polish and American archeologists unearthed what they believe to be some of the lectures halls of the library in May of 2004. They stated that the ruins they found contained thirteen lecture halls with seating for nearly five thousand students. That would mean that these grandiose lecture halls held an average of nearly four hundred students each, and in the center of each room was a large raised podium on which the instructor stood while lecturing. In 2005, some divers off the coast of Alexandria found what they believe to be another part of the library but much work remains to decipher exactly what they have found and what remains to be found.

Science had been born and reached great heights over the first thousand years. Through a combination of constant wars, large centralized government and religious intolerance, in societies that did not appreciate science, knowledge of our physical world faded away. Once this occurred, science remained almost unheard of in Europe for another thousand years.

However, the light of science did not ever completely extinguish and continued to burn like the embers of a snuffed fire for ten centuries. Moreover, just as in the birth of science, many factors played into its rebirth. Always thinking and discovering, the human mind never stops being curious.

Bibliography

About.com. “Map of Greece and Timeline.” 8 January 2005. <http://atheism.about.com/library/FAQs/religion/blgrk_greece04.htm>.

Arab, Sameh, M. Dr. “Bibliotheca Alexandria.” *The Ancient Library and the Re-Built* [sic] *of the Modern One*. 4 October 2005. <http://www.arabworldbooks.com/bibliothecaAlexandrina.htm>.

Benigni, U. “Catholic Encyclopedia.” *Capua*. English translation by Gerald M. Knight. 2005. <http://www.newadvent.org/cathen/03319a.htm>.

BiblePlaces.com. “Pergamum.” 1 January 2003. <http://www.bibleplaces.com/pergamum.htm>.

Bibliotheca Alexandrina. “The Ancient Library.” 2002. 30 December 2002. <http://www.bibalex.gov.eg/MainFrames.asp?LangID=1&Type=1&ID=1&Name=no%20text&FacilityID=77&IsAbstract=1&TemplateID=1>.

Bolling, George. “The Alexandrian Library.” 1907. Online edition 1999. 25 April 2001. <http://www.newadvent.org/cathen/01303a.htm>.

British Museum. “The British Museum Guide to Ancient Egypt.” New York: Thames and Hudson. 1992.

Burke, James. “The Day the Universe Changed.” Boston: Little, Brown and Company. 1985.

Cooperativa Sociale ‘Il Sogno'. “Romeguide.” 19 March 2005. <http://www.romeguide.it>.

Damascius. “Life of Isidore.” *The Life of Hypatia*. English translation by Jeremiah Reedy. 4 October 2005. <http://www.cosmopolis.com/alexandria/hypatia-bio-suda.html>.

de Solla Price, Derek. “An Ancient Greek Computer.” Scientific American. June 1959: 60-67. August 2005. <http://www.giant.net.au/users/rupert/kythera/kythera3.htm>.

Devlin, Keith. “The Math Gene.” Great Britain: Basic Books. 2000.

Diogenes Laertius. “Lives of Eminent Philosophers, V. II” (Books VII-IX). English translation by R.D. Hicks. London: Harvard University Press. 1979.

---. “Lives of Eminent Philosophers” (Books I-VI). English translation by C.D. Yonge. August-September 2005. <http://classicpersuasion.org/pw/diogenes/>.

Dodson, Aidan. “Monarchs of the Nile.” New York: The American University in Cairo Press. 2000.

Encyclopedia Britannica. “Democritus.” 27 April 2001. <http://www.britannica.com/eb/article?eu=30391>.

Finley, M.I. “The Ancient Economy.” Berkeley: University of California Press. 1973.

Freeman, Charles. “Egypt, Greece and Rome.” New York: Oxford University Press. 1996.

Grun, Bernard. “The Timetables of History.” New York: Simon and Schuster. 1991.

Hellenic Ministry of Culture. “Abdera.” 2001. 21 December 2002. <http://www.culture.gr/2/21/211/21119a/e211sa05.html>.

Hippocrates. “On Airs, Waters and Places.” Circa 400 BCE. English Translation by Francis Adams. Republished Online. MIT Internet Classics Archive. August 2005. <http://classics.mit.edu/Hippocrates/airwatpl.html>.

---. “On Ancient Medicine.” Circa 400 BCE. English Translation by Francis Adams. Republished Online. MIT Internet Classics Archive. August 2005. <http://classics.mit.edu/Hippocrates/ancimed.html>.

---. “On Injuries of the Head.” Circa 400 BCE. English Translation by Francis Adams. Republished Online. MIT Internet Classics Archive. August 2005. <http://classics.mit.edu/Hippocrates/headinjur.html>.

---. “The Law.” Circa 400 BCE. English Translation by Francis Adams. Republished Online. MIT Internet Classics Archive. August 2005. <http://classics.mit.edu/Hippocrates/headinjur.html>.

---. “The Oath.” Circa 400 BCE. English Translation by Francis Adams. Republished Online. MIT Internet Classics Archive. August 2005.

<http://classics.mit.edu/Hippocrates/headinjur.html>.

---. “On The Surgery.” Circa 400 BCE. English Translation by Francis Adams. Republished Online. MIT Internet Classics Archive. August 2005. <http://classics.mit.edu/Hippocrates/headinjur.html>.

History Channel, The. “Heron of Alexandria.” Broadcast 27 March 2005.

---. “Meteors: Fire in the Sky.” Broadcast 9 August 2005.

---. “Modern Marvels: Poisons.” Broadcast 8 October 2005.

---. “The Oracle of Delphi Secrets Revealed.” Broadcast 23 April 2005.

---. “The Rise and Fall of the Spartans, The.” Broadcast 22 March 2005.

---. “Search for Atlantis.” Broadcast 1 February 2005.

---. “Seven Wonders of the Ancient World.” Broadcast 1 February 2005.

---. “The True Story of Alexander the Great.” Broadcast 3 April 2005.

History Channel.com. Various. 2001-2005. <http://www.historychannel.com>.

John, Bishop of Nikiu. “Chronicle.” *The Life of Hypatia*. 4 October 2005. <http://www.cosmopolis.com/alexandria/hypatia-bio-john.html>

Matyszak, Phillip. “Chronicle of the Roman Republic.” New York: Thames and Hudson. 2003.

O'Connor, J.J. and E.F. Robertson. “Aristarchus.” 1999. 25 April 2001. <http://www-groups.dcs.st-Andrews.ac.uk/~history/Mathematicians/Aristarchus.html>.

---. “Callippus.” 11 January 2003. <http://www-gap.dcs.st-and.ac.uk/~history/Mathematicians/Callippus.html>.

---. “Democritus.” 1999. 25 April 2001. <http://www-groups.dcs.st-Andrews.ac.uk/~history/Mathematicians/Democritus.html>.

---. “Pythagoras of Samos.” 1999. 27 April 2001. <http://www-

groups.dcs.st-Andrews.ac.uk/~history/Mathematicians/Pythagoras.html>.

Plutarch. "Moralia, v. IV." English translation by F. C. Babbit. Cambridge, MA: Harvard University Press. 1993.

---. "Lives; Demosthenes and Cicero, Alexander and Caesar." English translation by Bernadotte Perrin. Cambridge, MA: Harvard University Press. 1994.

Roux, Georges. "Ancient Iraq." 3rd Edition. London: Penguin Press. 1992.

Sagan, Carl. "Cosmos." Episode 1. *The Shores of the Cosmic Ocean*. Atlanta, Georgia: Turner Home Entertainment. Carl Sagan Productions. VCR recording: 1989.

Science Channel, The. Hannibal: The Man who Hated Rome. Broadcast 24 March 2005.

---. Science Wonders; Gallo-Roman Secrets. Broadcast 14 April 2005.

---. Seven Wonders of Ancient Rome. Broadcast 15 March 2005.

---. What the Ancients Knew. Broadcast 15 March 2005.

---. Who Killed Julius Caesar? Broadcast 27 March 2005.

Seife, Charles. "Alpha & Omega." New York: Viking. 2003.

Smith, William. English translation by Bill Thayer. "A Dictionary of Greek and Roman Antiquities." 2005. <http://penelope.uchicago.edu/Thayer/E/Roman/Texts/secondary/SMIGRA/home.html>.

Socrates Scholasticus. "Ecclesiastical History". *The Life of Hypatia*. 4 October 2005. <http://www.cosmopolis.com/alexandria/hypatia-bio-socrates.html>.

Speller, Elizabeth. "Following Hadrian." New York: Oxford University Press. 2003.

Starr, Chester G. "A History of the Ancient World." New York: Oxford University Press. 1991.

Surintendance of Pompeii. "The Eruption of Vesuvius." 17 June 2005. <http://www2.pompeiisites.org/database/pompei/Pompei2.nsf/pagine/E4331B05D118

F066C1256ABC005A2C8A?OpenDocument>.

Taylor, Brian. Plato. 27 April 2001: <http://www.briantaylor.com/Plato.htm>.
Technology Museum of Thessaloniki. "Ctesibius of Alexandria." *Ancient Greek Scientists*. 18 September 2005:
<http://www.tmth.edu.gr/en/aet/1/31.html>.

Turizm.net. "Ionia and the Ionian Thinkers." 1998. 19 December 2002. <http://www.turizm.net/cities/ionia/index.html>

Week, The. *Wit & Wisdom*. 5.195. New York: The Week Publications. 18 Feb 2005.

Whitehouse, David, Dr. BBC News. "Library of Alexandria Discovered." 12 May 2004. 4 October 2005. <http://news.bbc.co.uk/1/hi/sci/tech/3707641.stm>.

Wikipedia. Various. 2005. <http://en.wikipedia.org/wiki/Main_Page>.

Wildberg, Christian, "John Philoponus", *The Stanford Encyclopedia of Philosophy (Fall 2003 Edition)* Edward N. Zalta (ed.). 26 September, 2005.
<http://plato.stanford.edu/archives/fall2003/entries/philoponus>.

Wolfe, Fredrick, Dr. "History of Science." Keene State College (NH) course. Attended Fall, 2001.

About the Author

This work is a life-long goal for the author. Ever since first becoming enamored with the Library of Alexandria from the Carl Sagan series "Cosmos," James Maynard has longed to write an entire book on the subject and how the lives of the scientists involved were shaped by and helped to shape, the world around them.

James Maynard has a Bachelor of Science in chemistry and physics, with a minor in history, from Keene State College in New Hampshire.

For three years, Maynard wrote "StarWatch," a weekly newspaper column on astronomy, which was published in New Hampshire and Vermont. He also taught the astronomy study sessions as a student instructor at Keene State College for three semesters and has guest lectured in classrooms and at various events throughout New England.

In the beginning of 2005, Maynard edited the second edition of "Phenomenal Physics" by Dr. Russell Harkay, his former faculty advisor. This work, nearly four hundred pages long, is an inquiry-based approach to physics, governing aspects of the science from kinematics to quantum theory, with simple do-it-yourself experiments and over one hundred diagrams. Several notable colleges have adopted or are currently considering adopting the work for use in their introductory physics courses. More information about the book (including how to purchase a copy, or to adopt it in your classroom) can be found at www.phenomenalphysics.com.

Currently, Maynard runs his own business, providing web design and author services, along with educational support throughout the state of New Hampshire and beyond.

Future planned projects include a possible second and third volumes to this book. The second book would cover the rebirth of science during the Renaissance, beginning with Leonardo Da Vinci and continue (as currently planned) through Charles Darwin (1809 - 1882). The third and final book in the trilogy will likely begin with James Clerk Maxwell (1831–79), who laid the groundwork for modern electrical theory and end in the modern day with such scientists as Brian Greene and Stephen Hawking.

James Maynard may be contacted through the website for this book, www.lightofalexandria.com.